Inventing the Indigenous

In the wake of expanding commercial voyages, many people in early modern Europe became curious about the plants and minerals around them and began to compile catalogs of them. Drawing on cultural, social, and environmental history, as well as the histories of science and medicine, this book shows how, amidst a growing reaction against exotic imports – whether medieval spices like cinnamon or new American arrivals like chocolate and tobacco – learned physicians began to urge their readers to discover their own "indigenous" natural worlds. In response, compilers of local inventories created numerous ways of itemizing nature, from local floras and regional mineralogies to efforts to write the natural histories of entire territories. Tracing the fate of such efforts, the book provides new insight into the historical trajectory of such key concepts as indigeneity and local knowledge.

Alix Cooper is Assistant Professor of History at the State University of New York at Stony Brook, where she teaches early modern European history and the histories of science, medicine, and the environment.

Inventing the Indigenous

Local Knowledge and Natural History in Early Modern Europe

ALIX COOPER
State University of New York,
Stony Brook

CAMBRIDGE
UNIVERSITY PRESS

CAMBRIDGE UNIVERSITY PRESS
Cambridge, New York, Melbourne, Madrid, Cape Town, Singapore,
São Paulo, Delhi, Dubai, Tokyo

Cambridge University Press
32 Avenue of the Americas, New York, NY 10013-2473, USA

www.cambridge.org
Information on this title: www.cambridge.org/9780521124010

First published 2007
This digitally printed version 2009

A catalog record for this publication is available from the British Library

Library of Congress Cataloging in Publication data
Cooper, Alix, 1966-
Inventing the indigenous : local knowledge and natural history in early
modern Europe / Alix Cooper.
p.   cm.
Includes bibliographical references and index.
ISBN-13: 978-0-521-87087-0 (hardback)
ISBN-10: 0-521-87087-9 (hardback)
1. Traditional medicine–Europe.   2. Traditional medicine–Europe–Sources.
3. Natural history–Europe.   4. Natural history–Europe–Sources.
5. Europe–History–1492-   I. Title.
GR880.C67   2007
940.2´1–dc22   2006037259

ISBN 978-0-521-87087-0 Hardback
ISBN 978-0-521-12401-0 Paperback

*To my parents
and to Tim*

Contents

Acknowledgments

I owe a great many debts to those who have helped me with this project. Let me thank first and foremost those who gave me support and advice early on: Everett Mendelsohn, who always found time to meet and talk, and lend an encouraging ear to my ideas; Katy Park, who read draft after draft and gave me some of the most lucid and helpful comments I have ever received; and Mario Biagioli, who counseled me on the parallels between the early modern court and modern academia. Several other individuals helped out so much they deserve special mention. Dorinda Outram tackled each chapter with bravery, fortitude, and vision. Lisbet Rausing helped steer the project as it got underway, with stimulating conversations over coffee and chocolate. And Paula Findlen, who saw some merit in an early paper, helped introduce me to a broader community of scholars of early modern history and natural history. More recently, I have benefited particularly from the assistance and insights of several additional individuals: Sara Lipton, who read an earlier version of the project cover to cover and gave sterling advice on its revision; Ned Landsman, who offered some extremely useful suggestions; and the anonymous readers of the manuscript for Cambridge University Press, with their own extremely helpful comments. My profound thanks to you all.

The History Department of the State University of New York at Stony Brook has provided an extremely congenial base for my work on this project, as have a number of other institutions over the years: the History of Science Department at Harvard University, the History and Environmental Studies Departments at the University of Puget Sound, and the Institut für Wissenschaftsgeschichte at the Universität Göttingen. Let me thank all of the faculty, staff, graduate students, and undergraduates with whom I've worked; I can't possibly list you all, but you know who you are. Audiences at numerous conference presentations and talks have also given me extremely stimulating feedback. I'd like to thank the following people in particular for advice, comments, and conversation: Martha Baldwin, Mason Barnett, David Blackbourn, Ann Blair, Peter Buck, Joan Cadden, Jane Camerini, Bill Clark, Hal Cook, Lorraine Daston, Margaret Garber, Jan Golinski, Bonnie Gordon, Anthony Grafton, Ernie Hamm, Steve Harris, Beth Hyde, Thomas Junker, Martina Kaup, Michael Kempe, Tomomi Kinukawa,

Sachiko Kusukawa, Manuel Lerdau, Mary Lindemann, Pam Long, Henry Lowood, Rob Martensen, Bruce Moran, Jean Dietz Moss, Martin Mulsow, Janice Neri, Bill Newman, Brian Ogilvie, Denise Phillips, Karen Reeds, Andrea Rusnock, Simon Schaffer, Thomas Schnalke, Londa Schiebinger, Anne and Jim Secord, Pamela Smith, Emma Spary, Govind Sreenivasan, Peter Stevens, Andre Wakefield, and Alice Walters. Many others have helped me as well, beyond my ability to document; if you don't find your name on this list, consider yourself included.

This book would not have been possible without the support of the National Science Foundation, which provided me with financial assistance early on. I was also fortunate to receive a fellowship from the Deutscher Akademischer Austauschdienst (DAAD), which gave me a year of research in Germany. Additional support from the Center for European Studies at Harvard, the Mellon Foundation, the Dibner Institute for the History of Science and Technology, the Max-Planck-Institut für Wissenschaftsgeschichte in Berlin, the Maria Sibylla Merian Postdoctoral Fellowship of the University of Erfurt, and the Princeton Library helped me to conduct further research and writing. Most recently, a National Endowment for the Humanities Fellowship granted me a year at the Folger Shakespeare Library in Washington, DC, where I was able to extend the comparative dimensions of my project, and to benefit from the fruitful companionship of Kay Edwards, Bill Sherman, Rebecca Laroche, Anston Bosman, Pat Parker, and Tim Raylor. I am extremely grateful to each of these institutions and granting agencies for their support.

I am likewise deeply indebted to the librarians and staff of the various collections I have visited over the course of my research. In the United States, I have drawn on the resources of the New York Public Library, the libraries of the New York Academy of Medicine and of the New York Botanical Garden, the Folger Shakespeare Library, the Library of Congress, the Smithsonian Institution Libraries, the National Library of Medicine, the National Agricultural Library, and the Johns Hopkins University Libraries, as well as many of the various libraries at Harvard University (in particular the Houghton Library, the Botany Libraries, Special Collections at the Museum of Comparative Zoology Library, the Kummel Geological Library, the Countway Medical Library, and the Andover-Harvard Theological Library). In Continental Europe, meanwhile, the Universitätsbibliothek Göttingen and its very helpful staff have provided me time and time again with an extremely hospitable base for research. I would also like to thank the staffs of several other libraries where I have spent considerable time, namely the Universitätsbibliothek Erlangen, the Herzog August Bibliothek in Wolfenbüttel, the Bayerische Staatsbibliothek in Munich, the Staatsbibliothek Preussischer Kulturbesitz in Berlin, the Universitäts- und Forschungsbibliothek Erfurt/Gotha, and the Zürich Zentralbibliothek. On the other side of the English Channel, I have over the years consulted the British Library and the libraries of the Wellcome

Acknowledgments

Institute, the Royal Society, the Linnaean Society, the
Natural History, and Kew Gardens. When all else has fa
to Donna Sammis, the incomparable Interlibrary Loan
Brook, whose ability to conjure up rare and obscure ma
at times little short of miraculous.

Earlier versions of some material from this book have
cle form, namely in "The Indigenous versus the Exotic: I
Origins in Early Modern Europe," *Landscape Research* 28
(available at http://www.tandf.co.uk) and "'The Possibilit
The Inventory of 'Natural Riches' in the Early Modern Gerr
in *Oeconomies in the Age of Newton*, ed. Margaret Scha
Marchi (Durham, NC: Duke University Press, 2003), 129–1
nally appeared as the annual supplement to volume 35 of the
of Political Economy. Likewise, part of chapter 2 appears in
Vernacular Worlds: Language, Nature, and the 'Indigenous' in
Europe," *Journal of East Asian Science, Technology, and M
coming (as part of a special issue on "Global Science and Co
tory: Jesuits, Science, and Philology in China and Europe, 155c
grateful to the representatives of these publications for permiss
those materials here.

Let me also thank Carolyn Edrich, Lydia Garcia, Seth G.
Krivicky, Katarina Wong, and Kate Blair-Dixon, without whi
might have remained mere scribbles; Eric Crahan, my editor;
Chaturvedi at Techbooks for all his patient help coordinating the
Finally, I'd like to give my deepest thanks to my parents and to
who have always been there for me through thick and thin. It's ti
true: the book could not have been completed without them.

List of Illustrations

Introduction

In the year 1643, on the shores of the Baltic, an obscure author published a small book on the plants to be found growing near his home town. Nicolaus Oelhafen's treatise was tiny, but it discussed what its author felt was a significant problem, one which extended far beyond its immediate setting, the merchant town of Danzig (today's Gdańsk). Why, Oelhafen complained, were so many people in his day fascinated by "strange" natural objects, "brought from faraway regions at great expense," while they "trod underfoot" those to be found at home? Rebuking them for their "ingratitude," he bitterly remarked that "Meanwhile, those things which grow under our own sun, in our own soil . . . if they don't lie entirely neglected and in contempt, are at any rate held to be viler than seaweed"![1] In his book, Oelhafen attempted to reintroduce his readers to the richness and variety of their own easily accessible countryside by compiling a detailed inventory of hundreds of local plant species, together with notes on where they could be found. By thus documenting local nature, he hoped, he could help to remedy his compatriots' ignorance while reestablishing a sort of balance and harmony in the greater world.

By taking this step, Oelhafen joined himself to a much larger enterprise. For across early modern Europe, many of his contemporaries – in such areas as Italy, France, England, the Netherlands, and the scattered territories of the Holy Roman Empire – were also beginning to contribute to "natural history," as they saw it, by documenting their own local natural worlds. Natural history, which comprised the study of rocks, plants, animals, and any other phenomena that might conceivably be described as "natural," was a pursuit with a venerable genealogy dating back to Greco-Roman antiquity.[2]

---

[1] Nicolaus Oelhafen, *Elenchus plantarum circa nobile Borussorum Dantiscum suâ sponte nascentium* (Danzig: typis & impensis Georgi Rheti, 1643), 1–2.

[2] See Nicholas Jardine, James Secord, and Emma Spary, eds., *Cultures of Natural History* (Cambridge: Cambridge University Press, 1996). Brian W. Ogilvie, *The Science of Describing: Natural History in Renaissance Europe* (Chicago: University of Chicago Press, 2006), 87–89, has recently argued that there is in fact a sharp discontinuity between ancient and Renaissance natural history, in other words that since no uninterrupted *community* of naturalists persisted throughout antiquity and the Middle Ages, the discipline itself must be seen as having been invented by Renaissance naturalists, who first established such an enduring community.

But Oelhafen's early modern counterparts had few words to describe exactly what it was that they were doing, in their efforts to investigate *local* nature in particular. The ubiquity of the term "local" is itself a relatively modern phenomenon; during the early modern period, it was used only in certain fairly narrow contexts, for example to discuss "local motion" in physics. Some compilers of inventories, then, declared the essence of their projects to be the study of their "domestic" natural worlds, while others talked of the "indigenous" or the "native," or used other similar terms (these decisions, of course, being highly dependent on the languages they spoke and wrote). Many just simply announced their intention to focus on natural objects in a particular place, whether a town or entire territory. Gradually, these compilers of inventories became aware of each other's existence; they began to cite each other and to compare their own local natural phenomena, wherever in Europe they might be, with those elsewhere. And gradually, they came to see their projects as sharing a common goal: not only the furthering of knowledge about the natural world in general, but also the furthering of a very specific sort of natural knowledge, that of "indigenous" natural kinds profoundly influenced by the places where they were to be found.

This book explores the meanings of the "indigenous" and related concepts in early modern Europe. When we use the term "indigenous" today, we tend to refer almost exclusively to the *non*-European – to those species, peoples, cultures, and knowledges most dramatically affected by the Columbian Encounter and its aftermath. Yet over the course of the early modern period, Europe saw the emergence of a fascination with a very different "indigenous": its own. Many early modern Europeans, as they struggled to make sense of the kinds of diversity they confronted from the fifteenth century on – previously unknown peoples, rediscovered ancient authorities, disturbing religious differences – sought new ways of understanding their worlds, and especially of coping with what they often perceived as "strange" and "foreign" influences.[3] In the process, many of them came to see these influences as embodied not just in human affairs, but also in the *material* world, most visibly in the trade in foreign medicines and exotic substances that had existed ever since antiquity, but had expanded substantially following medieval urbanization and the Columbian Encounter itself.[4] Debating

However, since early modern naturalists did in fact frequently draw on Pliny as a model, and the range of his concerns actually corresponds quite well with theirs, this book will use the term "natural history" in its broader chronological and thematic sense.

[3] Anthony Grafton, *New Worlds, Ancient Texts: The Power of Tradition and the Shock of Discovery* (Cambridge, MA: Harvard University Press, 1992).

[4] This topic has long been studied primarily by historians of medicine and pharmacy, as well as by economic historians and food historians. In recent decades, however, cultural historians have begun to contribute as well: see for example Wolfgang Schivelbusch, *Tastes of Paradise: A Social History of Spices, Stimulants, and Intoxicants*, translated by David

the qualities and merits of these substances, many early modern Europeans thus came to interpret their experiences of the foreign in large part through the natural world as well as the human one. And as they grappled with issues of geography, identity, and natural origins, many Europeans began to look *inwards* as well as outwards. In short, they began to pay attention to an "indigenous" located within Europe itself.

This may seem a controversial claim. But it is one rooted in intellectual debates and practices within early modern Europe, ones we have long since forgotten. In the wake of the Columbian voyages and early colonial endeavors overseas, a number of impulses joined to promote the scrutiny of local nature in Europe. New forms of fascination with the material world led to new conceptions of knowledge. New preoccupations with difference, both between and within communities, prompted new technologies for the gathering and recording of information. And polemics arose in which many Europeans – from physicians to popular pamphleteers – began to question the value of what they termed "exotic" substances more generally. Challenging boosters of expensive and fashionable remedies from afar, whether lavishly-prepared medicines long imported from the Mediterranean world or the increasingly trendy hot beverages of chocolate, coffee, and tea, some physicians in particular began, in reaction, to declare the need to take inventory of what they called the "indigenous" or "domestic" natural worlds of their own towns and territories. The resulting movement reached deep into Europe, attracting supporters not only in such colonial powers as England, France, and the Netherlands, but also, even more prominently, in the fragmented and decidedly non-colonial territories of the Holy Roman Empire, where local institutions and sentiments combined to produce the strongest push for the rediscovery of European natural objects and environments. In each of these places, people began to put pen to paper and to attempt, haltingly at first, to catalogue the "lowly" and "humble" weeds and pebbles in front of their doorsteps.

This book is thus, in part, about the ways in which, during the early modern period, the "indigenous" natural worlds of early modern Europe came to be debated and, ultimately, painstakingly documented. It was in Europe, rather than its colonies, that the kinds of works we today call "local floras" – books that catalogued the plant species to be found within a given radius of a town (often three, four, or five miles) – first began to be written. While medieval authors and, even more notably, the humanist botanists of the early Renaissance had shown a keen eye for local nature, their tendency

Jacobson (New York: Vintage Books, 1992); the articles in Roy Porter and Mikuláš Teich, eds., *Drugs and Narcotics in History* (Cambridge: Cambridge University Press, 1995); and David T. Courtwright, *Forces of Habit: Drugs and the Making of the Modern World* (Cambridge, MA: Harvard University Press, 2001).

had been to embed their descriptions of local species within universalizing works, ones which aimed to encompass all existing knowledge.[5] But early modern local florists gloried in their self-prescribed limitations to the local, explicitly restricting themselves to the pursuit of species "indigenous" or "native" to strictly limited regions. Such works were soon followed by other local inventories, from mineralogical surveys of areas' "subterranean riches," to ambitious schemes to write the "natural histories" of entire territories. The production of these kinds of inventories, which would ultimately shape many of the most basic structures and assumptions of today's environmental surveys, came to constitute one of the most significant arenas through which early modern Europeans engaged in reflecting on their own natural worlds – and, ultimately, on their perceptions of their own place within them.

By investigating this series of attempts to rediscover European nature, *Inventing the Indigenous* pursues several broader goals. One of these is to reconsider the ways in which Europeans thought about issues of geography and identity during this crucial period, so often labeled the "Age of Discovery." Recently, in the wake of the quincentenary of Columbus's first American voyage, a veritable explosion of scholarship on Europeans' encounters with extra-European peoples has occurred, examining these encounters anew from a wide range of critical perspectives, including those of postcolonialism and the emerging field of Atlantic history.[6] This literature has brought many new insights. For example, while some scholars of colonialism have unfortunately tended to treat Europe as a monolithic entity, others have begun to use more sophisticated analyses to reveal the ways in which religiously and politically diverse European polities in fact drew on colonial encounters to shape their identities in very different ways.[7] Similarly, studies of the ways

[5] Jerry Stannard, "Natural History," in *Science in the Middle Ages*, ed. David C. Lindberg (Chicago, IL: University of Chicago Press, 1978), 429–460; Karen Meier Reeds, "Renaissance Humanism and Botany," *Annals of Science* 33 (1976): 519–542; Karen Meier Reeds, *Botany in Medieval and Renaissance Universities* (New York: Garland, 1991); and Ogilvie, *The Science of Describing*.

[6] To cite just a few of the most prominent works belonging to this literature: Stephen Greenblatt, *Marvelous Possessions: The Wonder of the New World* (Chicago, IL: University of Chicago Press, 1991); Mary Louise Pratt, *Imperial Eyes: Travel Writing and Transculturation* (London: Routledge, 1992); Anthony Pagden, *European Encounters with the New World From Renaissance to Romanticism* (New Haven, CT: Yale University Press, 1993); and Stuart B. Schwartz, ed., *Implicit Understandings: Observing, Reporting, and Reflecting on the Encounters Between Europeans and Other Peoples in the Early Modern Era* (Cambridge: Cambridge University Press, 1994). On Atlantic history, see for example Bernard Bailyn, *Atlantic History: Concepts and Contours* (Cambridge, MA: Harvard University Press, 2005) and the essays in David Armitage and Michael J. Braddick, eds., *The British Atlantic World, 1500–1800* (Houndsmills, UK: Palgrave Macmillan, 2002), which both (despite the latter's geographical limits) include many useful references to broader work in the field.

[7] For a critique of some of the excesses of post-1992 revisionism, see Anthony Grafton, "The Rest versus the West," *New York Review of Books* 44, 6 (1997): 57–64, reprinted in *Bring Out Your Dead: The Past as Revelation* (Cambridge, MA: Harvard University Press, 2001),

in which differences between culturally, ethnically, and religiously disparate groups were perceived at the time have shown the complexity of early modern views on these differences, in an era when modern reifications of "race" had not yet fully developed.[8] In short, as recent research has revealed, contacts with newly-trafficked continents reached much more deeply into particular European societies than has previously been realized, as new ideas about their own place in a broader world subtly shaped their self-conceptions.[9]

Yet early modern Europeans grappled with issues of geography and identity not only through reports of new and strange peoples, but also – as scholars have only recently begun to recognize – through the *natural* world, both near and far. Europeans had long been accustomed to attaching meanings to natural objects based on their perceived origins, experiencing exotic products like spices, for example, as freighted with the mystery of the Eastern lands they came from, while viewing the vegetables that grew in peasants' gardens as emblematic of their "lowly" and humble nature.[10] This tendency

---

77–93. One key work demonstrating Europeans' highly diverse approaches to colonialism is Patricia Seed, *Ceremonies of Possession: Europe's Conquest of the New World, 1492–1640* (Cambridge: Cambridge University Press, 1996), though it too has been criticized as promoting a monolithic view of European societies, this time on a national level.

[8] Ivan Hannaford, *Race: The History of an Idea in the West* (Baltimore, MD: Johns Hopkins University Press, 1996); George M. Fredrickson, *Racism: A Short History* (Princeton, NJ: Princeton University Press, 2002); Joyce Chaplin, *Subject Matter: Technology, the Body, and Science on the Anglo-American Frontier, 1500–1676* (Cambridge, MA: Harvard University Press, 2001); Jorge Cañizares-Esguerra, "New World, New Stars: Patriotic Astrology and the Invention of Indian and Creole Bodies in Colonial Spanish America, 1600–1650," *American Historical Review* 104, 1 (1999): 33–68; and the special issue titled "Constructing Race: Differentiating Peoples in the Early Modern World," *William and Mary Quarterly*, 3rd ser., 54, 1 (1997).

[9] The classic work on the impact of the New World on the Old is J. H. Elliott, *The Old World and the New 1492–1650* (Cambridge: Cambridge University Press, 1970), though see also J. H. Elliott, "Final Reflections: The Old World and the New Revisited," in *America in European Consciousness, 1493–1750*, ed. Karen Ordahl Kupperman (Chapel Hill: University of North Carolina Press, 1995), 391–408. Recent studies complicating Elliott's theses include Grafton, *New Worlds, Ancient Texts*; Kathleen Wilson, *The Sense of the People: Politics, Culture, and Imperialism in England, 1715–1785* (Cambridge: Cambridge University Press, 1998); Kathleen Wilson, *The Island Race: Englishness, Empire, and Gender in the Eighteenth Century* (London: Routledge, 2002); and Benjamin Schmidt, *Innocence Abroad: The Dutch Imagination and the New World, 1570–1670* (Cambridge: Cambridge University Press, 2001). On the development not merely of new interests in the exotic, but of *exoticism* per se, see the recent work by Peter Mason, *Infelicities: Representations of the Exotic* (Baltimore, MD: Johns Hopkins University Press, 1998), and Benjamin Schmidt, "Inventing Exoticism: The Project of Dutch Geography and the Marketing of the World, circa 1700," in *Merchants and Marvels: Commerce, Science, and Art in Early Modern Europe*, ed. Pamela H. Smith and Paula Findlen (New York: Routledge, 2002), 347–369.

[10] On spices see Schivelbusch, *Tastes of Paradise*, 3–14; on vegetables, Paul Freedman, *Images of the Medieval Peasant* (Stanford, CA: Stanford University Press, 1999), 154 and Allen J. Grieco, "The Social Politics of Pre-Linnean Botanical Classification," *I Tatti Studies: Essays in the Renaissance* 4 (1992): 131–133.

seems only to have intensified in the wake of the Columbian voyages. As natural objects flowed in from an increasingly wide array of far-off continents, Europeans constructed imaginative geographies around their supposed origins. The New World medicaments known as "Brazil wood" and "balsam of Peru," for example, advertised their exotic genealogy through their very names, and descriptions of their virtues reflected this positioning.[11] Cartographers, meanwhile, drew strange creatures onto their new maps to fill uncharted spaces, and these came to symbolize entire continents; thus, for example, images of macaws, opossums, and armadillos increasingly took on the symbolic freight of South America in its entirety.[12] Not only plants and animals, but also a wide range of other kinds of natural phenomena were assigned their places in the European imagination. The appearance in 1494 of the disease now known as syphilis, for example, sparked a controversy around its own naming, as soldiers on the Italian battlefields where it first struck debated whether to call it the "French" or the "Neapolitan" disease.[13] This kind of imaginative geography was, obviously, often mistaken in its attributions of origin. The wild "Turkey" fowl brought back from the New World had, for instance, no connection whatsoever with the Ottoman Empire.[14] But early modern Europeans nevertheless seem to have found natural objects "good to think with," to paraphrase Lévi-Strauss.[15] Literally thousands of treatises were published over the course of the early modern period debating the merits of particular substances, from local beers or wines to exotic tinctures. In almost every case, the geographical origins of each item, as well as its prospects for replication or naturalization in

---

[11] For these examples, see Antonio Barrera-Osorio, *Experiencing Nature: The Spanish American Empire and the Early Scientific Revolution* (Austin: University of Texas Press, 2006). A thriving new literature has begun to emerge, based on the analysis of particular commodities in Atlantic or global contexts, and tracing their shifting cultural meanings: see for example John Brewer and Roy Porter, eds., *Consumption and the World of Goods* (London: Routledge, 1993); Arjun Appadurai, ed., *The Social Life of Things: Commodities in Cultural Perspective* (Cambridge: Cambridge University Press, 1996); Jordan Goodman, *Tobacco in History: The Cultures of Dependence* (London: Routledge, 1993); and Marcy Norton, *Sacred Gifts, Profane Pleasures: A History of Tobacco and Chocolate* (Ithaca, NY: Cornell University Press, forthcoming).

[12] Wilma George, *Animals and Maps* (Berkeley: University of California Press, 1969), 56–85.

[13] The former term stuck, to the indignation of Gallic physicians who repudiated the dubious honor. See Claude Quétel, *History of Syphilis*, translated by Judith Braddock and Brian Pike (Baltimore, MD: Johns Hopkins University Press, 1990), 10, and Jon Arrizabalaga, John Henderson, and Roger French, *The Great Pox: The French Disease in Renaissance Europe* (New Haven, CT: Yale University Press, 1997), 40.

[14] Ken Albala, *Eating Right in the Renaissance* (Berkeley: University of California Press, 2002), 233.

[15] For the original reference see Claude Lévi-Strauss, *Totemism*, translated by Rodney Needham (Boston: Beacon Press, 1963), 89. Another example of this phenomenon can be seen in the case of the court of Louis XIV, where certain kinds of flowers became powerful symbols for the king's own reign: see Elizabeth Hyde, *Cultivated Power: Flowers, Culture, and Politics in Early Modern France* (Philadelphia: University of Pennsylvania Press, 2005).

Europe, were presented as key topics for consideration. Natural objects thus offered Europeans attractive opportunities to think not only about faraway places, but also about where they themselves stood in a rapidly-changing world.

A striking example of this phenomenon may be seen in a cycle of four seventeenth-century paintings on the popular early modern theme of the "Allegory of the Continents," completed by the Flemish still-life master Jan van Kessel of Antwerp between the years 1664 and 1666. In these paintings, devoted to "Europe," "Asia," "Africa," and "America," respectively, van Kessel allegorized each continent as a queen, surrounded by a plethora of artifacts and, most prominently, natural objects clearly set forth as emblematic of the continent itself. Thus "Africa," for example, features a gigantic lion being stroked by its queen, while "America" is adorned by anteaters, an armadillo, a monkey, and several exotic birds.[16] Let us turn our attention, though, to the painting of Europe, or "Europa" as it is in fact titled (see Figure 1). Here Europe herself, represented as a queen, is seated in a large hall crammed full of objects and artifacts. Through a giant archway on the painting's left side can be seen the Castello Sant' Angelo with its bridge over the Tiber, placing the scene firmly in the traditional European cultural capital of Rome. Inside the hall, meanwhile, are ranged a vast array of both natural and artificial items which, it soon becomes apparent, symbolize the products of Europe. Among the artifacts shown strewn around the room, for example, are a celestial globe; several suits of medieval armor; a tall flag; assorted statues in the wall niches; an hourglass; a papal tiara; a portrait of Alexander VII (the Pope at the time); the Bible; and in the foreground, appearing somewhat incongruous amidst these more elevated objects, a set of playing cards and a tennis racket. Here, then, are depicted many of the most important symbols of European culture, representing its military, technological and scientific achievements as well as its religious triumphs, and not omitting its recreational pastimes – all displayed in liberal profusion around the figure of Europa herself.

Yet these symbols of European culture are in many ways overshadowed by the representations of European *nature* that occupy an even more prominent role in this painting. For the smiling queen's gaze is admiring *not* the above-mentioned symbols of her power, but rather a gigantic horn of plenty, stuffed full of fruit and grains, being handed to her by a cherub half its size. Meanwhile, at the very center of the picture stands a man (could he be Jan

[16] To add to the complexity of this cycle of paintings, Jan van Kessel placed a series of sixteen miniatures around the frame of each, surrounding its central panel and depicting animals and natural scenes associated with cities or places to be found on each continent; for the sake of simplicity, these are not treated here, though they reinforce many of the points made above. For further discussions of this cycle of paintings and of the broader genre of the "Allegory of the Continents," see Ulla Krempel, ed., *Jan van Kessel d. Ä., 1626–1679. Die Vier Erdteile* (Munich: Alte Pinakothek, 1973), and Sabine Poeschel, *Studien zur Ikonographie der Erdteile in der Kunst des 16.-18. Jahrhunderts* (Munich: Scaneg, 1985).

Figure 1. Jan van Kessel, *Europa* (central panel), 1664–1666. The crowned female
figure at the left represents Europe; the man at the center, possibly the artist himself.
Note the abundance of both natural and artificial objects symbolizing the wealth
of Europe. Courtesy of the Bayerische Staatsgemäldesammlungen, Alte Pinakothek
Munich.

van Kessel himself?) holding up and gesturing at a painting of butterflies,
dragonflies, and other insects, depicted flat against its surface as if pinned.
This painting is, in turn, surrounded by others: a gigantic still-life of car-
nations, roses, and tulips emerging from a tiny vase; another painting of
butterflies, this time portrayed in mid-flight against a background of peaceful hills; and, most curious of all, a painting depicting writhing snakes and
caterpillars spelling out the name of the artist himself. Nor is this cluster of
paintings, occupying a vast block of space in the image, the only reference to
the natural world. Shells spill out over the floor, perilously close to Europa's
and the cherub's feet, while an open book reveals still more images of butterflies and other insects, a closed book labeled "Plinius" alludes to the famous
Roman's *Natural History* (which indeed enjoyed a considerable revival during the early modern period), and in the lower left corner, yet another painting
(half-draped) can be seen, illustrating mandrake roots. Meanwhile, above all
this profusion, murals of marine invertebrates, high on the topmost walls,
overlook the scene. All of these *naturalia* are presented as emblematic of
the European continent, bountiful in its harvests of grain, surrounded by
the sea as well as mistress of it, and of a wide variety of technologies for
understanding and representing the beauties of the natural world. For the

European viewer, in short, every natural object in this and other similar visual and verbal descriptions of the world was replete with meaning. Nature's productions helped serve as means of interpreting the geography of a world in flux, where trade and travel increasingly connected Europeans with wider horizons, and forced them to attempt to construct their own sense of their place in the world.

Under these circumstances, as this image suggests, Europeans began to pay new *kinds* of attention to natural objects, as well as to "nature" in the abstract. Collectors, for example, drew on correspondence networks and personal ties to assemble vast quantities of unusual and rare natural objects, which they then showcased in their cabinets of curiosities or *Wunderkammer*, with walls, cupboards, and even ceilings hung with specimens and/or depictions of *naturalia*.[17] Painters and engravers, meanwhile, carefully studied particular items so as to produce such depictions, sharpening their skills at new naturalistic forms of representation in the process.[18] Goldsmiths and other artisans labored to transform select natural objects, such as giant conch shells, into magnificently-crafted artifacts like drinking cups, inviting these objects' users to reflect and converse on the paradoxical relationships between art and nature.[19] Courtiers at Renaissance princely palaces – and, eventually, the earliest scientific academies – honed their wits on discussions of striking natural phenomena from the mysterious "Bologna stone" to the "Medicean stars" observed by Galileo (now known as the moons of Jupiter).[20] And, last but not least, a wide range of writers and compilers scratched their heads and attempted to figure out how, using newly-arrived printing technologies and older manuscript ones, to develop new intellectual tools to enable them to set these newly vibrant natural worlds down on paper.[21]

[17] Oliver Impey and Arthur MacGregor, eds., *The Origins of Museums: The Cabinet of Curiosities in Sixteenth- and Seventeenth-Century Europe* (Oxford: Clarendon Press, 1985); Paula Findlen, *Possessing Nature: Museums, Collecting, and Scientific Culture in Early Modern Italy* (Berkeley: University of California Press, 1994).

[18] See for example Svetlana Alpers, *The Art of Describing: Dutch Art in the Seventeenth Century* (Chicago, IL: University of Chicago Press, 1983).

[19] Pamela H. Smith, *The Body of the Artisan: Art and Experience in the Scientific Revolution* (Chicago, IL: University of Chicago Press, 2004).

[20] Mario Biagioli, *Galileo, Courtier: The Practice of Science in the Culture of Absolutism* (Chicago, IL: University of Chicago Press, 1993); Bruce T. Moran, ed., *Patronage and Institutions: Science, Technology and Medicine at the European Court, 1500–1750* (Woodbridge: Boydell, 1991).

[21] See for example the articles in Marina Frasca-Spada and Nick Jardine, eds., *Books and the Sciences in History* (Cambridge: Cambridge University Press, 2000); Helmut Zedelmaier and Martin Mulsow, eds., *Die Praktiken der Gelehrsamkeit in der frühen Neuzeit* (Tübingen: Niemeyer, 2001); Mario Biagioli and Peter Galison, eds., *Scientific Authorship: Credit and Intellectual Property in Science* (New York: Routledge, 2003); and, though it deals with a slightly later period, Daniel R. Headrick, *When Information Came of Age: Technologies of Knowledge in the Age of Reason and Revolution, 1700–1850* (Oxford: Oxford University Press, 2000).

Historians of science in particular have, over the past several decades, done much to illuminate how these and similar practices came to infuse the study of nature in early modern Europe with still further cultural meaning and importance. Whereas traditional Aristotelian natural philosophy, as taught at medieval universities, had emphasized a "common-sense" understanding of nature, grounded on the commonly observed attributes of living organisms and other natural phenomena, the "new science" of sixteenth- and seventeenth-century Europe came to focus more and more on strange and unusual phenomena, on "particulars" and other isolated "facts" that often posed challenges to traditional natural-philosophical explanations.[22] Virtuosi sought out *naturalia* that were rare and unusual, that challenged conventional expectations of nature, and sought to explain them. Many of the natural objects that attracted the most interest within learned circles were, in fact, exotic. Stuffed birds or animals from the Indies, or depictions of strikingly-shaped or -colored fruits of the tropics, fascinated viewers through their revelations of the diversity of forms that nature could produce. Though few natural inquirers were willing to undertake perilous journeys to new continents themselves to collect strange specimens, they nonetheless hastened to examine them as they arrived in Europe, and avidly read accounts of natural phenomena from newly-trafficked lands, in search of whatever new insights about nature's workings these might provide.[23]

Yet as this book demonstrates, even while early modern Europeans sought out the rare and exotic, new and divergent ways of valuing nature simultaneously came into being, as many – especially the great majority of naturalists who would not have dreamt of travelling overseas – also began to pay new attention to what they called the "humble" and "common," even "vulgar" natural worlds surrounding them. The "rarities" of nature, they argued, could be found as well at home as abroad, and even the most apparently undistinguished kinds of plants or minerals might possess hidden value.[24] These kinds of objects, they felt, were well worth cataloguing in

22 Lorraine Daston, "Baconian Facts, Academic Civility, and the Prehistory of Objectivity," *Annals of Scholarship* 8 (1991): 337–363; Lorraine Daston and Katherine Park, *Wonders and the Order of Nature, 1150–1750* (New York: Zone Books, 1998).

23 For recent studies of early modern Europeans' scientific interests in the foreign, see the articles in Londa Schiebinger and Claudia Swan, eds., *Colonial Botany: Science, Commerce, and Politics in the Early Modern World* (Philadelphia: University of Pennsylvania Press, 2005); Pamela H. Smith and Paula Findlen, eds., *Merchants and Marvels: Commerce, Science, and Art in Early Modern Europe* (New York: Routledge, 2002); and, though it deals primarily with eighteenth-century developments, Londa Schiebinger, *Plants and Empire: Colonial Bioprospecting in the Atlantic World* (Cambridge, MA: Harvard University Press, 2004).

24 Keith Thomas, *Man and the Natural World: Changing Attitudes in England, 1500–1800* (London: Allen Lane, 1983), 58 and 66–69, discusses the long-standing application of traditional social hierarchies, like those of "nobility," to the nonhuman world, and the gradual eclipse during the early modern period of these distinctions; his broader argument about

their own right – hence the profusion of local floras, regional mineralogies, and other kinds of local natural histories that began to be produced documenting towns' and territories' natural "wealth." This development has, by and large, received little attention. On the whole, most historians – and, for that matter, other scholars in the humanities and social sciences – have long tended to regard topics relating to the natural world itself as beyond their purview.[25] Meanwhile, those scholars who *have* taken natural history seriously have, until quite recently, focused overwhelmingly on its classificatory aspects, to the exclusion of the many other meanings it held within early modern European culture.[26] Yet early modern writings on local nature, however obscure their "stay-at-home" authors, are, in fact, well worth our notice.[27] As they demonstrate, the early modern period saw the rise of new ways of valuing and understanding European objects and environments. By recovering this lost historical episode, and its consequences, this book aims to enhance significantly our understanding of how early modern Europeans actually thought about ideas of geography and identity, as they saw them mirrored in the natural world.

This book also has a broader goal, namely to reevaluate how we think today about issues of the local and of "local knowledge." In recent years, work in a variety of disciplines has come to draw heavily on these concepts. Anthropologist Clifford Geertz's influential *Local Knowledge*, for example, following on his earlier *Interpretation of Cultures*, exerted considerable

the early modern rise of sentimental, non-utilitarian attitudes towards nature has, however, been widely criticized. See also Grieco, 131–149.

[25] Recently, however, this has been changing, as historians of art and literature have taken an interest in representations of nature, and as environmental historians and commodity historians have sought to reintegrate natural environments and objects into history; for discussions of the latter two approaches, see William Cronon, "A Place for Stories: Nature, History, and Narrative," *Journal of American History* 78 (1992): 1347–1376, and the works cited in note 10 above.

[26] As a number of scholars have begun to show, the pursuit of early modern natural history in fact had many different goals, extending far beyond the merely taxonomic; see for example Jardine et al., eds., *Cultures of Natural History*; Paula Findlen, *Possessing Nature*; Londa Schiebinger, *Nature's Body: Gender in the Making of Modern Science* (Boston: Beacon Press, 1993); Lisbet Koerner, *Linnaeus: Nature and Nation* (Cambridge, MA: Harvard University Press, 1999); E. C. Spary, *Utopia's Garden: French Natural History from Old Regime to Revolution* (Chicago, IL: University of Chicago Press, 2000); David Freedberg, *The Eye of the Lynx: Galileo, His Friends, and the Beginnings of Modern Natural History* (Chicago, IL: University of Chicago Press, 2002); Schiebinger, *Plants and Empire*; and Ogilvie, *The Science of Describing*.

[27] In fact, by far the majority of Europeans were and have always been "stay-at-homes," in the sense that they contemplated broader worlds from the perspective of the armchair traveler; see for example Mary B. Campbell, *The Witness and the Other World: Exotic European Travel Writing, 400–1600* (Ithaca, NY: Cornell University Press, 1988) and Susanne Zantop, *Colonial Fantasies: Conquest, Family, and Nation in Precolonial Germany, 1770–1870* (Durham, NC: Duke University Press, 1997).

influence on scholars inspired by the book's implied project: namely the recovery and "thick description" of individuals' or communities' ways of knowing the world, previously ignored or suppressed in favor of seemingly universal "modern" forms of knowledge.[28] The concept of "local knowledge" has since been applied to a wide variety of situations and problems. Researchers in environmental studies and international development, for example, have used it repeatedly to explore contemporary controversies from England to the Philippines, showing how ordinary people in these places possessed crucial understandings and practices, for example related to agricultural sustainability or natural limits, that had been dismissed by the "experts."[29] Historians, meanwhile, have long been accustomed to paying close attention to particular local contexts. But recently, exciting work has begun to appear on how, in non-European situations in particular, "indigenous knowledges" either came to be assimilated into European knowledge systems, or were rejected for inclusion in them.[30] The results of this work have been quite rewarding, and have opened new avenues for scholarly research.

What this book aims to do, however, is examine the trajectory of "local knowledge" *within* Europe itself. For it was within Europe, during the early modern period, that numerous claims came to be made about the emergence of a new and seemingly "universal" form of knowledge, namely that which we nowadays associate with modern science. Over the past several decades, many scholars have come to argue that some of the most important features of this new "universal" knowledge in fact had their roots in the "local" settings in which they originated: in the laboratories of Robert Boyle, for example, and in his assistants' intimate knowledge of what it took to make an experiment actually "work."[31] Much of this scholarship

[28] Clifford Geertz, *Local Knowledge: Further Essays in Interpretive Anthropology* (New York: Basic Books, 1983). This book followed Geertz's equally influential *The Interpretation of Cultures: Selected Essays* (New York: Basic Books, 1973).

[29] See for example Frank Fischer, *Citizens, Experts, and the Environment: The Politics of Local Knowledge* (Durham, NC: Duke University Press, 2000); and Alan Bicker, Paul Sillitoe, and Johan Pottier, eds., *Investigating Local Knowledge: New Directions, New Approaches* (Aldershot, UK: Ashgate, 2004).

[30] On the appropriation (or lack thereof) of indigenous knowledge during different periods, see Richard Grove, *Green Imperialism: Science, Colonial Expansion and the Emergence of Global Environmentalism, 1660–1880* (Cambridge: Cambridge University Press, 1994), 73–94; Judith Carney, *Black Rice: The African Origins of Rice Cultivation in the Americas* (Cambridge, MA: Harvard University Press, 2001); Jorge Cañizares-Esguerra, *How to Write the History of the New World: Histories, Epistemologies, and Identities in the Eighteenth-Century Atlantic World* (Stanford, CA: Stanford University Press, 2001); Schiebinger, *Plants and Empire*. On indigenousness itself, or "indigeneity," see Jace Weaver, "Indigenousness and Indigeneity," in *A Companion to Postcolonial Studies*, ed. Henry Schwarz and Sangeeta Ray (Oxford: Blackwell, 2000), 221–235.

[31] Steven Shapin and Simon Schaffer, *Leviathan and the Air-Pump: Hobbes, Boyle, and the Experimental Life* (Princeton, NJ: Princeton University Press, 1985).

has concerned itself with disciplines such as physics, whose laws are now indeed seen as applicable everywhere and in all situations. By probing the origins of these most paradigmatic of the modern sciences, and seeking to show the ways in which even the most apparently universal kinds of modern knowledge were shaped by the circumstances in which they were formed, researchers have been able to study the complex negotiations by which these kinds of knowledge first came to be accepted as authoritative, and ultimately universal.[32]

Natural history, however, presented a very different case, as early modern Europeans became more and more aware of the variety of forms that natural phenomena displayed across Europe and across the globe.[33] Because the species and natural objects they investigated tended to differ from place to place, very few naturalists were ever thus in a position to make "universal" claims, as did natural philosophers such as Boyle; instead, the vast majority of naturalists came, in their geographically-defined catalogues, to define themselves through their very possession of what we would nowadays call "local" rather than universal knowledge. Their efforts to do so were complicated by the intricate and often-shifting political geography of Europe during the Age of Religious Wars, and even afterwards. Many scholars have characterized this era before the emergence of the modern nation-state as one of intense localism and regionalism; though often seen as having been strongest in parts of Europe which, like Italy and the German territories, lacked centralizing monarchies, this is also frequently acknowledged as one of the basic conditions of premodern life throughout the entire continent,

---

[32] Examples of works along these lines would be too numerous to cite; for some recent theoretical approaches to this set of problems, see Adi Ophir and Steven Shapin, "The Place of Knowledge: A Methodological Survey," *Science in Context*, 4 (1991): 3–21; Steven Shapin, "Placing the View from Nowhere: Historical and Sociological Problems in the Location of Science," *Transactions of the Institute of British Geographers*, 23 (1998): 5–12; David N. Livingstone, *Putting Science in its Place: Geographies of Scientific Knowledge* (Chicago, IL: University of Chicago Press, 2003); and the articles in David N. Livingstone and Charles W. J. Withers, eds., *Geography and Revolution* (Chicago, IL: University of Chicago Press, 2005).

[33] Martin Rudwick has over his career repeatedly drawn attention to the ways in which the physical location of and difference between natural objects has shaped directions of scientific research; see for example his *The Great Devonian Controversy: The Shaping of Scientific Knowledge Among Gentlemanly Specialists* (Chicago, IL: University of Chicago Press, 1985); for a discussion of natural history specifically as a "science of difference," see Schiebinger, *Nature's Body*. See also Dorinda Outram, "New Spaces in Natural History," in N. Jardine, J. A. Secord, and E. C. Spary, eds., *Cultures of Natural History* (Cambridge: Cambridge University Press, 1996), 249–265, and Peter Dear's recent point about the direct relevance of geography to the making of natural-historical knowledge in particular in his "Space, Revolution, and Science," in David N. Livingstone and Charles W. J. Withers, eds., *Geography and Revolution* (Chicago, IL: University of Chicago Press, 2005), 38–39.

and it was one that naturalists had to adapt to as they attempted to discover and define "indigenous" nature.[34]

This localism or regionalism might take place on many different geographical levels. First, there was that of the town or village, which depending on its size might be walled off from the countryside for protection; the surrounding countryside, though, was often conceptualized as belonging to the town in question.[35] A person's primary political allegiance would often be to this most basic unit, as seen in the fact that references to a person's *patria* (literally "fatherland") frequently referred just to his or her town, nothing more; in early modern Europe, people from other towns were thus often considered "foreigners." Then there might follow a whole sequence of larger regional units of administration, depending on the political structure of the area in question; England, for example, inherited from the medieval period a complex pattern of boroughs, shires, and counties, while in the Holy Roman Empire, where over 1000 territories of different sizes effectively ruled themselves, these kinds of regional units ranged from tiny Church-owned landholdings to enormous princely states. Finally, on what we would nowadays call the "national" level, monarchies and, less commonly, republics or confederations held sway, if indeed their influence was felt at all. Many residents of territories within the Holy Roman Empire, for example, seem to have been barely aware of its presence, since it had far less impact on daily life than the more immediate contexts of town and territory; and to have conceived of themselves not so much as imperial subjects, but rather as members of a German "nation" defined more on ethnic and linguistic terms than political ones. When naturalists set out to write about "indigenous" or "domestic" natural worlds, then, they had many different options as to what might be the appropriate scale on which to explore.

In the process of writing their works, though, and thus constructing their own forms of what we might call "local" knowledge, naturalists were confronted with the existence of *other* forms of understanding of local natural worlds. For various other forms of writing about the local existed at the time, some of which did include reference to the "natural." For example, medieval city chronicles, in their narrations of natural disasters and urban response,

---

[34] See for example Dietrich Gerhard, "Regionalismus und ständisches Wesen als ein Grundthema europäischer Geschichte," *Historische Zeitschrift*, 174 (1952): 303–337; on the case of the German territories in particular, see Hajo Holborn, *A History of Modern Germany* (Princeton, NJ: Princeton University Press, 1982), I, 12–14 and II, 37–38; and Walter Bruford, *Germany in the Eighteenth Century* (Cambridge: Cambridge University Press, 1952), especially Ch. 1, "Kleinstaaterei." While recent works have urged the "provincializing" of Europe – i.e. Dipesh Chakrabarty, *Provincializing Europe: Postcolonial Thought and Historical Difference* (Princeton, NJ: Princeton University Press, 2000) – by examining developments in other parts of the world, an alternative strategy for doing this is to explore the ways in which Europe was already provincialized from within.

[35] Since extremely few urban areas in early modern Europe had more than 10,000 inhabitants, even those labeled "cities" at the time would nowadays be considered towns.

had long told stories about nature, while setting it into its human context. Renaissance humanists, meanwhile, both Italian and Northern, had come to compose a wide variety of works, from civil history to chorography (regional geography), in praise of their towns and territories; these had tended to intermix civil and natural history while rooting them in the classical genres of *laus* (praise) and panegyric.[36] Meanwhile, agents of European states had since the late medieval period begun, extremely gradually, to survey their territories in various ways, whether qualitative or quantitative, verbal or visual: from charging officials with updating or generating new records for tax purposes, for example, to commissioning maps of particular areas.[37] The state, however, was far from the only entity interested in making use of local knowledge. Cartographers, for example, found their knowledge of geographical techniques highly sought after by numerous groups in addition to the state, such as merchants eagerly seeking trade advantages, and printer-publishers well-aware of the commercial potential of the aesthetically pleasing new maps.[38] Meanwhile, still other forms of local knowledge about nature existed, from herbalists' familiarity with local plant remedies,[39] to

[36] On humanist local history and topography, see for example Hans Baron's classic *Crisis of the Early Italian Renaissance: Civic Humanism and Republican Liberty in an Age of Classicism and Tyranny* (Princeton, NJ: Princeton University Press, 1966); D. R. Woolf, *The Social Circulation of the Past: English Historical Culture, 1500–1730* (Oxford: Oxford University Press, 2003); Stan A. E. Mendyk, *'Speculum Britanniae': Regional Study, Antiquarianism, and Science in Britain to 1700* (Toronto: University of Toronto Press, 1989); Frank L. Borchardt, *German Antiquity in Renaissance Myth* (Baltimore, MD: Johns Hopkins University Press, 1971); and Gerald Strauss, *Sixteenth-Century Germany: Its Topography and Topographers* (Madison: University of Wisconsin Press, 1959).

[37] See for example Jacques Revel, "Knowledge of the Territory," *Science in Context*, 4 (1991): 133–161; David Buisseret, ed., *Monarchs, Ministers and Maps: The Emergence of Cartography as a Tool of Government in Early Modern Europe* (Chicago, IL: University of Chicago Press, 1992); Michel Foucault, "Questions on Geography," in *Power-Knowledge: Selected Interviews and Other Writings 1972–1977*, ed. Colin Gordon (Brighton, UK: Harvester, 1980), 74–7; and James C. Scott, "Nature and Space," in *Seeing Like A State* (New Haven, CT: Yale University Press, 1998), 11–52. Scott argues that these state-based projects had the effect of simplifying complex local conditions and making them "legible" to the state; for a cautionary note, though, see Chandra Mukerji, "The Great Forestry Survey of 1669–1671: The Use of Archives for Political Reform," *Social Studies of Science*, forthcoming, which argues that state agents, on the contrary, often cultivated "near-sightedness," producing accounts rich in local detail.

[38] On early modern mapping and its multiple audiences, see for example Jerry Brotton, *Trading Territories: Mapping the Early Modern World* (Ithaca, NY: Cornell University Press, 1997); Lesley Cormack, *Charting an Empire: Geography at the English Universities, 1580–1620* (Chicago, IL: University of Chicago Press, 1997); and Denis E. Cosgrove, *Apollo's Eye: A Cartographic Genealogy of the Earth in the Western Imagination* (Baltimore, MD: Johns Hopkins University Press, 2001).

[39] Because it was rarely recorded, whether for reasons of secrecy or illiteracy, historians have found the knowledge of herbalists on the village level very difficult to track (Renaissance "herbals," though they may have drawn on herbalists' experiences, were in most cases compiled by learned physicians or printers), but for some suggestive approaches to reconstructing this knowledge, see Jole Agrimi and Chiara Crisciani, "Immagini e ruoli della 'vetula' tra

peasants' understandings of the land they plowed.[40] During the early modern period, however, certain particular forms of local inventory – for example, the local floras and other genres that will be discussed in this book – came to acquire a special status among natural inquirers as privileged repositories of empirical information about the individual objects and species that made up the natural world. Over the course of this transformation, the authors of local natural inventories drew on and incorporated many different kinds of knowledge and expertise. Yet owing to the very "local" nature of these inventories, they ultimately came to be challenged as insufficiently "universal" for a world transformed in many ways by the new sciences.[41] One of the contributions of this book will be, hopefully, to illuminate some of these tensions and contradictions in light of their historical origins.

Let me reframe this set of problems in a somewhat different way: through the example of an intriguing map I found in the early pages of a standard botanical reference work, D. G. Frodin's *Guide to Standard Floras of the World*.[42] This "five-grade map of the approximate state of world floristic knowledge as of 1979" (see Figure 2) graphically displays a striking unevenness, or disparity, in modern botanists' perceptions of how much they know about different parts of the world. While the map shows much of the globe – Africa, Asia, and the Americas, for example – as lightly shaded, i.e. relatively poorly known, much of northern Europe stands out in contrast, densely colored, as quite well-studied. Some regions appear particularly dark, suggesting especially intense investigation: England, the Netherlands, Switzerland, much of Scandinavia, and a large swathe of central Europe. Here so much detailed information has apparently become available, in forms that professional botanists can digest, that certain small areas of the map are shown almost black with information.

This map presents us with a world which is, in large part, the outcome of the historical processes analyzed in this book. In this world, natural

sapere medico e antropologia religiosa (secoli XIII–XV)," in *Poteri carismatici e informali: chiesa e societa medioevali*, ed. Jole Agrimi (Palermo: Sallerio, 1992), 224–261; J. Burnby, "The Herb Women of the London Markets," *Pharmaceutical Historian* 13 (1983): 5–6; and Martha Baldwin, "Expanding the Therapeutic Canon: Learned Medicine Listens to Folk Medicine," in James Van Horn Melton, ed., *Cultures of Communication from Reformation to Enlightenment: Constructing Publics in the Early Modern German Lands* (Aldershot, UK: Ashgate, 2002).

[40] See Piero Camporesi, "Retarded Knowledge," in *The Anatomy of the Senses: Natural Symbols in Medieval and Early Modern Italy*, translated by Allan Cameron (Cambridge: Polity Press, 1994).

[41] On rejection of the "particular," even amidst the use of "particulars," in universal science, see Lorraine Daston, "How Nature Became the Other: Anthropomorphism and Anthropocentrism in Early Modern Natural Philosophy," in *Biology as Society, Society as Biology: Metaphors*, ed. Sabine Maasen (Dordrecht: Kluwer, 1995), 38–39.

[42] D. G. Frodin, *Guide to Standard Floras of the World*, 2nd ed. (Cambridge: Cambridge University Press, 2001), 12.

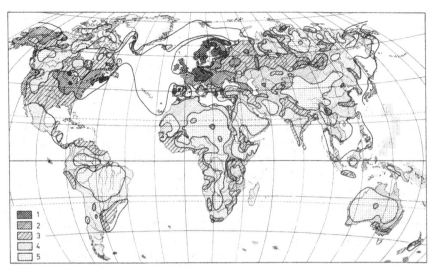

Figure 2. "Five-grade map of the approximate state of world floristic knowledge as of 1979." The more darkly an area is shaded, the more thoroughly it is believed to be scientifically "known." Contrast in particular the shading of various parts of northern Europe with that of Latin America, Africa, and Asia. From D. G. Frodin, *Guide to Standard Floras of the World*, 2nd ed. (Cambridge: Cambridge University Press, 2001), p. 12, as revised from E. J. Jäger in *Progress in Botany* 38 (1976), p. 317.

knowledge is far from evenly distributed. Even though an enormous quantity of books, articles, and other documents have come to be published over the past five centuries on the local natural worlds of particular places, certain areas have come to receive much more "scientific" attention than others, in ways which, as the global patterning of this particular map suggests, clearly go beyond mere random distribution. As the respective colorations of Europe and the rest of the world attest, historical conditions have here played an enormous role in shaping what has been judged or defined as "known," and what is seen as remaining "unknown." For example, the local flora has clearly come to serve, in this map, as an arbiter of "world floristic knowledge"; in other words, as the unit upon which, according to the map, botanical knowledge is to be built. *Inventing the Indigenous* explores how it was that the local flora, and its related genres, came to assume this role, despite their own "local" origins in early modern Europe; in short, how these kinds of judgments about the validity and extent of different kinds of knowledge came to be made in the first place. In the course of examining early modern European decisions to record and make public information about specific natural objects, the book thus simultaneously explores how it was that some crucial activities of science came to be distributed in unequal ways, and the broader implications of this development.

The chapters in the book move, both chronologically and thematically, from early concerns with the "indigenous" to their realization in various situations and contexts. Chapter 1 uses the tools of cultural and intellectual history to investigate the origins of early modern Europeans' interests in their own "indigenous" or "domestic" natural worlds, as they called them, amidst a new era of cross-cultural encounter and exchange. Looking at examples ranging from the early writings of Paracelsus, the controversial sixteenth-century medical reformer, through an assortment of herbal, medical, and travel writings of the next two centuries, the chapter traces how the very categories of the "indigenous" and the "exotic" (which came to be seen as the polar opposite of the "indigenous") came to be formed in the early modern European imagination.

Chapter 2, meanwhile, approaches these issues from a social and institutional perspective. It focuses on the emergence of one particular way of documenting the European "indigenous": namely through what we now call "local floras," or catalogues of plants to be found growing in a given area, which came to be compiled and published in ever-growing numbers throughout Europe. Investigating the origins and development of this genre in provincial towns, universities, and gardens, the chapter explores how the pursuit of "indigenous" nature came to be translated into practice, in the quest for a natural diversity to be found within Europe itself.

Turning to a different form of interest in the "indigenous," Chapter 3 investigates efforts in the German territories, after the end of the Thirty Years' War, to survey the full range of those territories' rocks and minerals. The resulting documents, which might be termed regional mineralogies, diverged from local floras in significant ways. Taking as a case study the striking example of the "lying stones" of Würzburg, and drawing on insights from the correspondence of late seventeenth- and early eighteenth-century writers of regional mineralogies, the chapter analyzes authors' attempts to publicize their own local areas' "natural riches" for the benefit of local economies and states, and argues that for inhabitants of early modern Central Europe, local rocks and minerals came to provide a way to discuss the "nature" of their own territories.

Chapter 4 explores the ways in which, in seventeenth- and eighteenth-century Europe, various individuals embarked on utopian attempts to describe the full "natural history" of entire regions – not only their plants or minerals but also a wide range of phenomena from their birds and bugs to their weather, climate, and landscape more generally. As this chapter shows, naturalists in different areas of Europe had very different approaches toward this goal. The chapter focuses on the efforts of naturalists in different areas to influence each other through the new international networks provided by scientific academies and journals, and on the problems they encountered in attempting to communicate their "local" knowledge to an increasingly far-flung audience.

Chapter 5, finally, takes stock of these various European efforts to document the "indigenous" in the natural world. Examining the very different ways in which the Swiss naturalist Johann Jakob Scheuchzer and his renowned Swedish counterpart Linnaeus framed their early-eighteenth-century histories and bibliographies of natural history, the chapter argues that Linnaeus's contemptuous dismissal of earlier local naturalists has masked the ways in which Linnaeans ended up incorporating many aspects of previous traditions of local natural history. The chapter argues that debates about "local knowledge" thus ended up being embedded within the genre of the inventory of nature itself.

Before launching into the book proper, however, a note on methodology may be helpful. Owing to the nature of the developments discussed, which unfolded over the course of centuries and in various different areas within Europe, the book will of necessity range rather broadly in both space and time. The approach will therefore be transnational, with the aim of contrasting the varying styles of local natural history that emerged in different places within Europe; however, considerable attention will be granted to Central Europe and in particular the early modern German territories, where, for reasons the book will attempt to explore, some of the earliest forms of local natural history originated.[43] Because of this emphasis on the origins of those forms of local natural history that still survive today, most of which did emerge within the geographical limits of the European continent, the book will therefore unavoidably be Eurocentric in its structure, i.e. colonial developments will be discussed only insofar as they influenced or were influenced by what happened in Europe. Despite this limitation, however, the book will indeed attempt to demonstrate some of the various ways in which broader interconnections between different parts of the world did indeed affect local natural history within Europe itself, as Europeans increasingly began to look outwards – though often, as we shall see, to the Old World rather than the New.[44] No single force or set of forces will be seen to emerge as a sole cause for the new focus on the indigenous in early modern Europe. Rather, numerous paths joined; this book will explore some of their interconnections.

---

[43] When the phrase "German territories" is used, it will normally refer to the territories within the early modern Holy Roman Empire (*Heiliges Römisches Reich deutscher Nation*); these generally, but not always, overlap with the German-speaking areas of Central Europe. If the terms "German" or "Germany" are used, they should be read as referring to the cultural context of both above-mentioned areas, not to the political unit of unified Germany as created in the nineteenth and twentieth centuries.

[44] On this issue more generally, see Jerry Bentley, *Old World Encounters: Cross-Cultural Contacts and Exchanges in Pre-Modern Times* (Oxford: Oxford University Press, 1993). For recent discussions of Atlantic vs. Old World encounters, see Joyce E. Chaplin, "Expansion and Exceptionalism in Early American History," *Journal of American History* 89 (2003): 1431–1455; Peter A. Coclanis, "*Drang Nach Osten*: Bernard Bailyn, the World-Island, and the Idea of Atlantic History," *Journal of World History* 13 (2002): 169–182.

Finally, a note on nomenclature may also be helpful. As has already been mentioned, the analysis of the very terminology that was and is still used to discuss the "local," "indigenous," and so forth will form one of the main tasks of this book. In general, my tendency will be to use the "actors' categories" whenever possible, that is to say the actual terms people used during the early modern period (or their closest English equivalents). However, this is not always possible, and in such cases, I will simply use words according to my best estimate of common English usage today. For example, when the word "local" is used outside of quotation marks, I will use it in its most common colloquial modern sense, namely as referring to phenomena occurring on a very small scale (say, within a village, or a small group of people); I will reserve the terms "regional" and "national" to refer to those larger scales of interaction. Likewise, I will refer to the genre of local plant inventories as the "local flora," since this is the term by which it has come to be known today; and, by extension, I will refer to similar genres that emerged in the early modern period as works of "local natural history," since no better umbrella term suggests itself. Similarly, even though few of the individuals I discuss would have thought of themselves primarily as "Europeans" – other forms of regional, religious, and ethnic identity would have been far more meaningful to them – I will indeed have to use this term at times, to contrast them implicitly or explicitly with non-Europeans. In general, however, my guiding principle will be to use and explain early modern categories whenever possible, and to attempt to make them meaningful to the modern reader; this is also, of course, the goal of the book as a whole.

1

Home and the World:
Debating Indigenous Nature

In 1652, London apothecary Nicholas Culpeper published an herbal titled *The English Physitian, or, An Astrologo-Physical Discourse of the Vulgar Herbs of this Nation*. This work, which became so popular that it was reprinted numerous times and, as "Culpeper's Herbal," remains in print even today, had a simple goal: to discuss how anyone could "cure himself, being sick, for three pence charge, with such things only as grow in *England*, they being most fit for English Bodies." Previous authors on plants, argued Culpeper, had failed to do this. Instead, he charged, they had "intermixed many, nay very many outlandish Herbs," and in the process caused immeasurable harm. He meant his book to remedy the situation.[1]

What did Culpeper mean by "outlandish Herbs," and why did he set out to write about "such things only as grow in *England*" in response? To answer these questions, it will be necessary to venture into a set of wide-ranging early modern debates over nature and the native, debates which had their origins long before Culpeper first set pen to paper. For when Culpeper used the word "outlandish," he did not mean merely to say that the herbs he was writing about were strange, perhaps, or weird – though he would almost certainly have appreciated these semi-pejorative connotations, and hoped that his readers would as well. But he was also using the word in its original, literal sense, now obsolete or almost so. "Outlandish" things were those that came from outside one's native land or country, that were foreign or exotic, native to another place, not one's own. According to Culpeper, foreign herbs were not only expensive, difficult to procure, and often adulterated, they were also not even "fit for English Bodies." What his countrymen needed, claimed Culpeper, was a source of information on those plants they could call their own. And in his *English Physitian*, he set out to do just that.

Culpeper was not alone. Over the course of the sixteenth and seventeenth centuries, a new enthusiasm for exploring the local natural worlds of early modern Europe came into being. This interest manifested itself in numerous ways, but especially in the pursuit of natural history, the time-honored

---

[1] Nicholas Culpeper, *The English Physitian: or, An Astrologo-Physical Discourse of the Vulgar Herbs of this Nation* (London: printed by Peter Cole, 1652), sig. A2v (emphasis in original).

study of such natural phenomena as plants and minerals. Amidst an age of exploration and long-distance travel, which saw the spread of European contacts worldwide, many Europeans came to busy themselves much closer to home, producing a wide range of works documenting the natural variety to be found not abroad, but rather within Europe itself. Herbals like Culpeper's proliferated, as did new genres like local floras or plant catalogues, regional mineralogies, and natural histories of entire territories. Each of these genres developed its own distinctive set of techniques for taking inventory of what came to be called the "indigenous," "native," or "domestic" natural productions of Europe. Drawing on a fierce set of polemics about natural origins, authors of these works came to contrast "indigenous" European natural objects with "exotic" imports from other lands. Even as boosters of these imports made lavish claims about their curative powers and other striking qualities, authors of the new local natural histories blasted exotic substances for their perceived moral and medical as well as economic dangers. In the process, the natural world came to serve as a mirror for concerns about geography and identity in early modern Europe.

This chapter will explore the origins of this early modern fascination with the "indigenous," and the debates that accompanied it. As the chapter will show, sixteenth- and seventeenth-century Europe saw the convergence of a number of factors that spurred individuals to begin to look in new ways at their local natural worlds, and to consider how these might be related to natural worlds elsewhere. Renaissance discoveries and rediscoveries, anxieties over the potentially corrupting effects of commerce, and tensions over hierarchies within and beyond the healing professions were all among the contexts that shaped the emergence of a powerful series of concerns about the European "indigenous" vis-à-vis the exotic. Taking one especially revealing early treatise as a case study, the chapter maps the general contours of the debate over the indigenous, before proceeding to examine how it unfolded over the course of the sixteenth century and on into the seventeenth. In this investigation into concepts and origins, no single force or set of forces emerges as a sole cause for early modern Europe's fascination with the indigenous. Rather, numerous paths joined; this chapter will explore some of their interconnections.

"THERE ARE IN GERMANY SO MANY MORE AND BETTER MEDICINES..."

Thus claimed Paracelsus, the notorious Swiss medical reformer and religious radical of the early sixteenth century, in his *Herbarius*, or treatise "concerning the Powers of the Herbs, Roots, Seeds, etc. of the Native Land and Realm of Germany." And this forms a perfect place to begin a discussion of sixteenth-century views on nature and its geography. Though the iconoclastic Paracelsus can hardly stand for a typical early modern individual (if such a person indeed ever existed!), nonetheless, his case provides considerable insight into broader sixteenth-century European concerns over the geographical origins of natural objects, concerns which would be expressed again and again by authors from a wide range of European polities over the course of the century. Paracelsus is justifiably better known for his medico-chemical and theological writings, in which he rejected the traditional scholastic learning of the medieval university and set forth his own idiosyncratic views on nature, medicine, and the occult realm; such writings comprise by far the greatest part of the vast body of work he produced.[2] But in his *Herbarius*, a short and fragmentary work composed in German during the 1520s but published only later in the century, we can nonetheless find a particularly striking vantage point into much broader sixteenth-century debates about the origins of natural objects, and the cultural meanings they acquired as a result of these origins.[3]

Paracelsus chose to commence his *Herbarius* with an appeal to his German-speaking readers. It is worth reproducing here, for it encapsulates many of the concerns he and other polemicists were to articulate:

Because I see that the medicines of the German nation come from far-off lands at great cost and with much care, effort, and travail, I have been moved to ask whether Germany might not itself be in command of medicines, and whether, without the foreign sort, these may exist also in its own domain.[4]

Several different themes may be discerned amidst Paracelsus's tangled phrasing. First, we observe a sort of "German" patriotism or cultural pride in Paracelsus's effort to prove his own region's remedies equal to (or better than) those of other areas, and to rescue them from their perceived neglect. This is a position resonant not only with the values of Renaissance civic humanism, with its focus on local pride, but also with the kinds of patriotic sentiments stirred up by the Reformation in northern Europe. As historians of this period have noted, Luther's appeals to the "Germans" and to the "German nation" were crucial in fostering a sense of regional identity in the fragmented territories of the Holy Roman Empire, long before

---

[2] Walter Pagel, *Paracelsus: An Introduction to Philosophical Medicine in the Era of the Renaissance* (Basel: Karger, 1958).

[3] This treatise has been translated into English in Bruce T. Moran, "The *Herbarius* of Paracelsus," *Pharmacy in History* 35 (1993): 99–127. For convenience, all quotations will be from this translation. The original German text is available as "Herbarius Theophrasti [Paracelsi] de virtutibus herbarum, radicum seminum etc Alemaniae, patriae et imperii" in *Theophrast von Hohenheim gen. Paracelsus Sämtliche Werke. I. Abteilung: Medizinische, naturwissenschaftliche und philosophische Schriften*, ed. Karl Sudhoff (Munich & Berlin: Oldenbourg, 1930), vol. II, 3–58. Moran, 101, explains the complicated publishing history of this treatise; although it was composed during the mid- to late 1520s, pieces of it only originally appeared separately, in 1568, and were first published together as an ensemble in 1570.

[4] Moran, 104; Sudhoff, ed., II, 3.

"nationalism" in its modern sense entered European discourse.[5] Paracelsus may thus be seen to echo Luther's vision of the "German nation" as a cultural unit, if not yet a political one. What is especially striking here, however, is Paracelsus's invocation of cultural pride in the context not only of ethnicity, but also of medicine, and thus of *nature*. For Paracelsus as well as many other writers, as we shall see, the geographic origin of natural substances would prove crucial in determining their status. This passage sets forth a theme – the appeal for people to investigate the natural products of their own areas – that would come to be a characteristic feature of writings on local natural history.

Another point of importance emerges from this passage, however. Paracelsus did not mention the "German nation" in isolation; rather, he explicitly contrasted its status with that of "far-off lands," and its medicines with those of "the foreign sort."[6] Here, in this set of rhetorical contrasts, we see Paracelsus's thoughts on nature in Germany as crucially linked to his opinions of "foreign" nature. In the conceptual opposition he set up between the "German" and the "foreign" appears the kernel of a dichotomy between the "indigenous" and the "exotic," one that would come to assume a major role in sixteenth-century writing on nature as well as on health. In Paracelsus's usage, this dichotomy may be crudely drawn, but it is vehement: "They want to prepare medicines from across the seas, when there are better remedies to be found in front of their own noses, in their own gardens. Look, dear readers, how contradictory it is that one can see so far that he sees across the ocean, but fails to see what is in the earth right in front of him."[7] For Paracelsus, the polarity he established between medicines from "across the seas" and ones "in the earth right in front" of a person was thus one heavily charged with meaning on a variety of levels, reaching beyond specific questions about specific drugs to encompass broader issues of the origins of knowledge itself.[8]

[5] See for example A. G. Dickens, *The German Nation and Martin Luther* (London: Edward Arnold, 1974), 1–48. On the fraught question of the emergence of nationalism more generally, see Benedict Anderson, *Imagined Communities: Reflections on the Origin and Spread of Nationalism* (London: Verso, 1991).

[6] For a further discussion of this kind of rhetorical strategy, see Orest Ranum, "Counter-Identities of Western European Nations in the Early-Modern Period: Definitions and Points of Departure," in *Concepts of National Identity: An Interdisciplinary Dialogue / Interdisziplinäre Betrachtungen zur Frage der nationalen Identität*, ed. Peter Boerner (Baden-Baden: Nomos, 1986), 63–78.

[7] Moran, 104; Sudhoff, ed., II, 4.

[8] That this was not an isolated theme, even in Paracelsus's own writings, may be seen in his well-known pamphlets on the treatment of syphilis, which sharply criticized the imported remedy of guaiac bark (shipped in from South America), while touting the curative powers of mercury, one of his famous chemical remedies: see *Vom Holtz Guaiaco gründlicher Heylung* (Nuremberg: Friedrich Peypus, 1529) and *Von der frantzösischen Kranckheit. Drey Bücher* (Nuremberg: Friedrich Peypus, 1530).

In order to understand Paracelsus's stance better, it is necessary to consider, as he did, the state and ambitions of early modern natural history, and in particular of botany, that branch of natural history devoted to accounting for the world of plants. Paracelsus roundly condemned previous efforts in this area: "Several German writers have come forth and have described herbs and plants in books. But their work is like the coat of a beggar, patched together from all sorts of things. But the whole is really nothing and it falls apart like a beggar's coat which can no longer stand its own weight; and so there is nothing there when one most needs it."[9] The tradition which Paracelsus thus slandered, calling its authors "raving sorts, these seducers, false inform-ers, and teachers of medicine,"[10] was a venerable one. Medieval compila-tions of *materia medica*, or medicinal substances, dated back to Dioscorides in the first century, often drawing their illustrations (correspondingly poor in quality, due to the passage of time) from the same purported sources; even after the advent of printing, these materials continued to be recycled. Though medieval people were, of course, far from blind to the natural world around them – as witnessed, to name just one example, by monk Walahfrid Strabo's delightful poetic exposition of his garden on Lake Constance in the ninth century[11] – nonetheless more traditional collections of remedies predominated, gathered not primarily from personal experience but rather from previous such collections.[12] Paracelsus certainly had obvious reasons for spurning these kinds of works, whose status as obvious compilations from different times and places might seem an apt target for his metaphor of the beggar's coat, "patched together from all sorts of things." Even dur-ing the Middle Ages, many of these works had already come under heavy criticism for their eclectic and often seemingly indiscriminate character.[13]

Yet it is worth noting that Paracelsus, in this passage, seems to single out "German" authors for particular opprobrium. Why? An answer may have to do with the general character of the herbals that, since the European adoption of moveable type half a century earlier, had begun to roll off the new printing presses in Mainz and other cradles of printing. We have seen Paracelsus's interest in identifying specifically "German" remedies; but the bulk of these early printed herbals, were, in fact, generally universalist in out-look and compass. Far from focusing specifically on the plants of any one area, most tended instead, in semi-encyclopedic fashion, to incorporate as much material as possible within their covers, merging all information into

[9] Moran, 105; Sudhoff, ed., II, 5.   [10] Moran, 105; Sudhoff, ed., II, 5–6.
[11] Walahfrid Strabo, *Hortulus*, translated by Raef Payne, with commentary by Wilfrid Blunt (Pittsburgh, PA: The Hunt Botanical Library, 1966).
[12] Jerry Stannard, "Natural History," in *Science in the Middle Ages*, ed. David C. Lindberg (Chicago, IL: University of Chicago Press, 1978), 429–460.
[13] Chiara Crisciani, "History, Novelty, and Progress in Scholastic Medicine," *Osiris* 6 (1990): 136.

a broader stream of plant lore cobbled together since antiquity.[14] Though compilers of these early herbals sometimes remarked on plants from particular places, in general they saw little need to stress or emphasize the local as such, but rather sought to accommodate all entries within a common framework. Inclusiveness was the general ethos; in these compilations, popular garden herbs and plants growing wild were juxtaposed, in ways often seemingly random, with rare and expensive tropical spices from thousands of miles away. It is perhaps this "patchwork" quality of the early herbals, the recycling of texts and images from a mix of previous sources, that Paracelsus most disliked. In his condemnation of their German compilers may be seen a frustration with the inclusive frameworks within which they (and all other such authors at the time) worked, that encouraged them to extend the fabric of an already existing "patchwork" of knowledge, rather than starting anew in some more radical or coherent way, in keeping with his reformist passions.

Yet it must be pointed out that attention to local plants did indeed come to be a key feature of the researches that went into producing sixteenth-century botanical works, including many herbals, and that some efforts in this direction had already begun by the time of Paracelsus's writing. There are several reasons why this was the case. Most importantly, those humanists who undertook the task of editing ancient botanical texts had found startling discrepancies between the plants described there, and the confusingly different set of plants they found around them. This problem was especially obvious for northern humanists in the German territories, England, the Netherlands, and Scandinavia, who, despite their best efforts, often could find little correspondence between Mediterranean floras and their own. Attempts to resolve these discrepancies through recourse to the time-honored method of scholarly conciliation – seeking to show that apparently discrepant phenomena were, in fact, one and the same – required research not only into the classical texts, but into the particulars of the modern plants themselves.[15] The critical attention paid by humanist botanists to their texts thus inspired an

---

[14] See Agnes Arber, *Herbals: Their Origin and Evolution: A Chapter in the History of Botany 1470–1670*, 3rd ed. (Cambridge: Cambridge University Press, 1986), 13–37. Arber, 20–22, does note one exception, the so-called *Latin Herbarius* (1484) which contained primarily plants native to or naturalized in the German territories; she believes it too, though, to be a medieval compilation.

[15] Karen Meier Reeds, "Renaissance Humanism and Botany," *Annals of Science* 33 (1976): 519–542; Karen Meier Reeds, *Botany in Medieval and Renaissance Universities* (New York: Garland, 1991); and Brian W. Ogilvie, *The Science of Describing: Natural History in Renaissance Europe, 1490–1620* (Chicago, IL: University of Chicago Press, 2006). On the relationship between humanism and science more generally, see Anthony Grafton, *Defenders of the Text: The Traditions of Scholarship in an Age of Science, 1450–1800* (Cambridge, MA: Harvard University Press, 1991); Ann Blair, *The Theater of Nature: Jean Bodin and Renaissance Science* (Princeton, NJ: Princeton University Press, 1997); and Gianna Pomata and Nancy Siraisi, eds., *Historia: Empiricism and Erudition in Early Modern Europe* (Cambridge, MA: The MIT Press, 2005).

equal attention to the examination and identification of the natural world around them, as revealed in the many local plants they uncovered.[16] Discarding the Dioscoridean tradition of medieval plant illustration, entrepreneurial botanists from the 1530s onwards took care to have new illustrations drawn from "life," often using local plants as models. And indeed it is the realism and accuracy of these oft-reproduced illustrations for which their volumes are today best known.[17] The challenge northern botany presented to classical traditions of botany stemmed in great part from these "learned empiricist" habits of local investigation, ones which would come to form a model more generally for later naturalists.

Paracelsus's blanket condemnation of botanical authors, however, was not confined to those who in his view had ignored true "German medicine." To trace his argument, it will again be necessary to quote at length:

Moreover, there are in Germany so many more and better medicines than are to be found in Arabia, Chaldaea, Persia, and Greece that it would be more reasonable for the peoples of such places to get their medicines from us Germans, than for us to receive medicines from them. Indeed, these medicines are so good, that neither Italy, France, nor any other realm can boast of better ones. That this has not come to light for such a long time is the fault of Italy, the mother of ignorance and inexperience. For the Italians saw to it that the Germans thought nothing of their own plants, but rather took everything from Italy itself or from beyond the sea . . .[18]

Here we see Paracelsus again drawing a number of conceptual oppositions: this time not merely between "German" plants and "foreign" ones, but more broadly between Germany and Italy, between northern and southern Europe, between Protestant and Catholic, and between Europe and the Islamic world. As we have seen above, several of these oppositions may be traced to the arguments of the northern humanists, while others (the contrast between Germany and "Arabia, Chaldaea, Persia, and Greece," for example) were clearly far more deeply rooted in a lengthy tradition of Old World encounters than in any New World ones.[19] What is most important to note here, however, amidst this proliferation of polarities, is Paracelsus's very reliance

[16] Paula Findlen, *Possessing Nature: Museums, Collecting, and Scientific Culture in Early Modern Italy* (Berkeley: University of California Press, 1994), 163–170, 179–184.

[17] Elizabeth Eisenstein, *The Printing Press as an Agent of Change: Communication and Cultural Transformations in Early Modern Europe* (Cambridge: Cambridge University Press, 1979), 265–267, 484–488; William Ivins, *Prints and Visual Communication* (Cambridge, MA: The MIT Press, 1967), 40–46; and Sachiko Kusukawa, "Leonhart Fuchs on the Importance of Pictures," *Journal of the History of Ideas* 58 (1997): 403–427.

[18] Moran, 104; Sudhoff, ed., II, 3.

[19] For the gradual formation of distinctions between European and non-European worlds, see Denys Hay, *Europe: The Emergence of an Idea* (Edinburgh: Edinburgh University Press, 1957); Kevin Wilson and Jan van der Dussen, eds., *The History of the Idea of Europe* (London: Routledge, 1995); and Anthony Pagden, ed., *The Idea of Europe from Antiquity to the European Union* (Cambridge: Cambridge University Press, 2002). As is evident from

on such polarities as a source of explanation. In the intellectual landscape of the early sixteenth century, unsettled by new continents and new ideas (such as the disturbing new divide between Catholic and Protestant Europe), such polarities served as a tool for unraveling identity, for attempting to define where one stood in relation to the rest of a world that had grown larger. Paracelsus used them to explain the course of medical and botanical history, and to justify his interest in one specific part thereof; others during the same period were to use similar constructs as frameworks for the writing of local histories or topographies, for example, or for the consolidation of power within increasingly confessionalized and territorial states. Later naturalists would frequently simplify Paracelsus's varied and multiple phrasings into a somewhat more consistent and homogeneous opposition between the "indigenous" and the "exotic"; Paracelsus himself, though, seems to have reveled in the opportunity here to let fly at a wide range of potential targets.

But Paracelsus's advocacy of "German" medicaments, and the venom he hurled at exotic remedies in his diatribes, must also be seen as borrowing from – as well as helping to regenerate – a specifically *medical* set of polemics. These drew on a lengthy tradition of debate over hierarchies within the healing professions. One of the main sources of conflict was the precarious relationship between university-educated physicians and guild-based apothecaries, over whom physicians frequently claimed supervisory authority. Tensions between the two groups often erupted in the context of debates about the authenticity and efficacy of particular remedies. Physicians, for example, often accused apothecaries of adulterating expensive foreign ingredients with cheaper local ones, the better to make an easy profit.[20] During the Middle Ages, apothecaries had frequently doubled as "spicers" and "pepperers," selling a wide variety of substances – often from abroad – for both medicinal and culinary uses. Druggists' continuing associations with expensive but easily-counterfeited foreign wares, and their frequent willingness to help patients evade doctors' fees by performing their own diagnoses and recommending their own remedies, made them an easy target for physicians' wrath. Commercial and economic concerns had thus long joined with professional ones to create tensions over the sale of exotic ingredients.[21]

the example above, although the term "Europe" did exist at the time, many other concepts were far more important in shaping identity.

[20] Edward Kremers and George Urdang, *History of Pharmacy*, 4th ed. (Madison, WI: American Institute of the History of Pharmacy, 1976), 20, 27–28, 33–34, and 69–72.

[21] See R. S. Roberts, "The Early History of the Import of Drugs into Britain," in *The Evolution of Pharmacy in Britain*, ed. F. N. L. Poynter (London: Pitman Medical Publishing Company Ltd., 1965), 165–185; Nancy G. Siraisi, *Medieval and Early Renaissance Medicine: An Introduction to Knowledge and Practice* (Chicago, IL: University of Chicago Press, 1990), 146–147; and Mary Lindemann, *Medicine and Society in Early Modern Europe* (Cambridge: Cambridge University Press, 1999), 89, 216. Just as sixteenth-century astronomers were reluctant to "deal in novelties," then – see Jean Dietz Moss, *Novelties in the Heavens: Rhetoric*

In the early sixteenth century, however, the advent of medical human-ism worsened relations between physicians and apothecaries even further. In the aftermath of the Byzantine Empire's collapse and the arrival of a flood of Greek-speaking refugees in Italy, humanist physicians lauded newly-received Greek editions of ancient medical authors. Simultaneously, though, they mounted harsh attacks on the medieval medicine of their predeces-sors, which they saw as having been corrupted by its extensive borrow-ings from the Arabic medical tradition. This corruption, medical humanists believed, had occurred in two main ways. First, they charged, Arabic writings had encouraged widespread medieval reliance on polypharmacy, the use of "compound" medicines made from multiple ingredients; in their stead, med-ical humanists advocated a return to the "simples" (remedies derived from just one plant, animal, or mineral) supposedly favored by the ancients.[22] Secondly, humanists complained, Arabic medical compendia had introduced into European practice a wide range of substances which, originating in the deserts of Arabia or even further afield, were far too "hot" and "spicy" for cooler European bodies accustomed to more northern climes; this, they warned, could lead to medically-dangerous overheating.[23] Ignoring the fact that such classical authorities as Galen and Dioscorides had themselves fre-quently recommended both compound medicines *and* Eastern spices in their own works,[24] humanist physicians condemned not only Arabic authors but also European pharmacists as degrading the classical legacy, as well as public health, through their continued advocacy of expensive and exotic drugs.

Far from taking sides in the quarrel between physicians and apothecaries, Paracelsus attacked both – indeed, all – parties. Not only did he decry the physicians who themselves prescribed questionable medicines from afar, he also denounced all of those who made the drugs in question available,

and Science in the Copernican Controversy (Chicago, IL: Chicago University Press, 1993) – so too learned medical practitioners felt they had every right to be suspicious of "exotics" which, growing far away, they could not examine for themselves in their places of origin. I am grateful to Jean Dietz Moss for this point; see also Ogilvie, 229–258.

[22] Owsei Temkin, *Galenism: Rise and Decline of a Medical Philosophy* (Ithaca, NY: Cornell University Press, 1973), 126–128.

[23] Heinrich Schipperges, "Der Anti-Arabismus in Humanismus und Renaissance," in *Ideologie und Historiographie des Arabismus* (Wiesbaden: Steiner, 1961), 14–25; see also Nancy G. Siraisi, *Avicenna in Renaissance Italy: The Canon and Medical Teaching in Italian Univer-sities after 1500* (Princeton, NJ: Princeton University Press, 1987), 66–76.

[24] Vivian Nutton, "The Drug Trade in Antiquity," *Journal of the Royal Society of Medicine* 78 (1985): 142–143; John Scarborough, "Roman Pharmacy and the Eastern Drug Trade: Some Problems as Illustrated by the Example of Aloe," *Pharmacy in History* 24 (1982): 135. On the recovery of Dioscorides more generally, see John M. Riddle, *Dioscorides on Pharmacy and Medicine* (Austin: University of Texas Press, 1985) and Richard Palmer, "Medical Botany in Northern Italy in the Renaissance," *Journal of the Royal Society of Medicine,* 78 (1985): 149–157.

from pharmacists to shippers and other intermediaries. He complained, for example, about "the deception of merchants, shopkeepers, and sellers of medicine, for these bring nothing pure to us from foreign shores. The same is also true of those middlemen who store up medicines and then resell them.... Those who carry medicines into German lands and seek their own profit from unsuspecting buyers are just as bad, and thus this stale merchant's treasure has gone bad and is altogether worthless by the time one delivers it to the person who is sick."[25] Paracelsus's attacks were thus not limited to any one group, such as physicians or apothecaries, but rather included all those who trafficked in the exotic in any way. The only medical practitioners whom Paracelsus exempted from his blanket condemnation were those who relied on "experience and personal practice," the method through which he claimed to have attained his own insights.[26] Paracelsus touted the experience gained by peasants and other inhabitants of the land, who, unencumbered by book-learning, often discovered local herbs of genuine efficacy.[27] His promotion of "German" herbs and rejection of their exotic counterparts thus bore strong links to his stance on authority in medicine, as well as within early modern society more generally.

One further strand of medical polemic in which Paracelsus engaged remains to be teased out. This has to do with his conception of the environment and its role in illness. For Paracelsus, as for Hippocrates many centuries before, and for the majority of early modern physicians, it was common knowledge that there was an intimate connection between geography and the diseases of the inhabitants of a given area.[28] Hippocrates and his early followers had focused their attention on individual Greek cities, and on colonial sites in the Mediterranean where Greek cities might potentially be founded, noting the correlations between low, swampy land and various fevers, and, on the other hand, higher, drier lands and relative health. Paracelsus, however, drew on an expanded notion of medical geography popular during the early modern period, one based not only on the correlation of place and illness, but of illness and cure. He held that local diseases had their

---

[25] Moran, 105; Sudhoff, ed., II, 4.   [26] Moran, 105; Sudhoff, ed., II, 5.

[27] Charles Webster, "Paracelsus: Medicine as Popular Protest," in *Medicine and the Reformation*, ed. Ole Peter Grell and Andrew Cunningham (London: Routledge, 1993), 70. Paracelsus also acknowledged the role of "old women" in providing herbals on a village level, though his assessment of them was more mixed. See Andrew Wear, *Knowledge and Practice in English Medicine, 1550–1680* (Cambridge: Cambridge University Press, 2000), 57, footnote 28.

[28] See Wesley D. Smith, *The Hippocratic Tradition* (Ithaca, NY: Cornell University Press, 1979); Caroline Hannaway, "Environment and Miasmata," in *Companion Encyclopedia of the History of Medicine*, ed. W. F. Bynum and Roy Porter (London: Routledge, 1993), 292–308; Genevieve Miller, "'Airs, Waters, and Places' in History," *Journal of the History of Medicine and Allied Sciences* 8 (1962): 129–140; and Conevery Bolton Valenčius, "Histories of Medical Geography," in *Medical Geography in Historical Perspective*, ed. Nicolaas A. Rupke (London: Wellcome Trust Centre for the History of Medicine at UCL, 2000), 3–28.

own local remedies: "Each land, to be sure, gives birth to its own special kind of sickness, its own medicine, and its own physician."[29] His system thus had no place for "exotic" remedies for local problems. Even chemicals, after all, could be synthesized at home, in the basement laboratories he urged all adepts and seekers-after-knowledge to build. In this view, his admiration of indigenous plants thus conflicted not at all with his often-repeated advocacy of chemical medicines; both approaches represented ways for knowledge to be taken out of the hands of physicians and entrusted to those who, pursuing the path of "experience," might genuinely turn the light of nature to the human good.

This section has focused on Paracelsus not so much because of his general notoriety, but rather because of the frank and explicit manner in which he set forth his opinions on the moral economy of the natural world. These offer clues to the attraction of local nature for Paracelsus. In the *Herbarius*, he articulated an antipathy toward foreign medicines, together with a strong approval of what he called "German" ones, in a way that prefigured later polarizations of the "indigenous" and the "exotic" in the natural world. Though in the *Herbarius* these categories are not explicitly labeled, they are nonetheless invested with the emotional intensity characteristic of the Reformation era in general. While Paracelsus's commitment to indigenous remedies is consistent with his other writings, he did not set down in any greater detail his thoughts on native nature. Nor can the scope of his polemical ambition in this treatise, the entirety of the "Native Land and Realm of Germany," be said to have been a realistic target for comprehensive coverage at the time; even though some later authors were to try writing explicitly German, French, or English herbals, they came nowhere close to making complete surveys of their respective territories. When local floras began to be written early in the seventeenth century, their authors were to choose much more limited areas of inquiry, such as the area around a single city or town. Such severely restricted outlooks were alien to Paracelsus. In the *Herbarius* itself, he declared that the ideal physician would have to travel broadly, as he himself had: "I would think that German doctors, in so far as they might wish to be highly regarded as widely experienced, would have considered that one should become a wanderer *(ein perambulanus)* so as to learn and experience things first hand."[30] For despite what he claimed in this treatise, Paracelsus was in fact *not* comfortable with any truly "localist" outlook. In his writings, he repeatedly urged the physician to read the "book of nature," with each country visited representing one more page turned.[31] And

---

[29] Moran, 104; Sudhoff, ed., II, 4.   [30] Moran, 104–5; Sudhoff, ed., II, 4.

[31] See "The Fourth Defense: Concerning My Journeys" in Paracelsus's *Sieben Defensiones, Verantwortung über etliche Verunglimpfungen seiner Missgönner*, translated into English in *Paracelsus: Four Treatises*, ed. Henry E. Sigerist (Baltimore, MD: Johns Hopkins University Press, 1941), 29; see also Sudhoff, ed., XI, 145–146.

Paracelsus's *Herbarius*, at least in its published version, ended up including an extended version of the medical merits of coral, neither a plant *nor* local in any way to the Germanies, but possessing, he claimed, astonishing virtues in the treatment of a wide range of medical conditions.[32] For Paracelsus, as for other sixteenth-century writers, however much they might come to praise local nature, it did not suffice to satisfy their ultimate ambitions.

"GARLIC AND ONIONS"

In his advocacy of local remedies, and in his decision to locate these in opposition to those from "foreign" lands, Paracelsus was far from unique. Over the course of the sixteenth century, the kinds of contrasts he made between "German" and "foreign" natural products were expressed by numerous other writers and incorporated into a more general polarity: one most frequently expressed as that of the "indigenous" versus the "exotic." This development may be seen clearly, for example, in the herbals and other botanical works published during the remainder of the sixteenth century. Numerous universalist compendia continued to be printed, showcasing their breadth and comprehensiveness in their inclusion of a rapidly increasing number of species, from New World and Old. But they came more and more over the course of the century to divide their subject conceptually into the two distinct and opposed components, one domestic, the other foreign. The exact labels used differed from region to region and language to language – those writing in the scholarly language of Latin overwhelmingly referred to the "indigenous" and "exotic," for example, while Germans usually contrasted the *einheimisch* with the *ausländisch*, and the English preferred to speak of the "native" vis-à-vis the "alien," "outlandish," "exotick," or simply "strange"[33] – but the basic dichotomy remained the same. Thus the title of a book might promise to supply "not only the indigenous, but also the exotic," or vice versa, with the two categories presented as complementary; or, alternatively, a book might be advertised as focusing just on one or the other. And when the first local floras and similar localist inventories of nature came to be written, early in the seventeenth century, it was this category of the "indigenous," also referred to as the "domestic" – and almost always used to refer to the *European* – that would come to define their subject matter, in opposition to the category of the "exotic." The "indigenous" thus arose as

---

[32] Moran, 119–123; Sudhoff, ed., II, 40–46.

[33] See the entries on these and similar terms, for example, in J. A. Simpson and E. S. C. Weiner, eds., *The Oxford English Dictionary* (Oxford: Oxford University Press, 1989), I, 314–6; IV, 944–945; V, 551–552; VI, 51–52; VII, 867; X, 235–238 & 1020–1021; XVI, 841–845; and Jacob and Wilhelm Grimm, eds., *Deutsches Wörterbuch* (Leipzig: S. Hirzel, 1854–1971), I, cols. 900–1; III, cols. 197–198; IV, cols. 125–129.

one half of a matching pair, through which European natural objects were directly linked to and contrasted with those elsewhere.

None of these various labels or patterns of thought was, of course, entirely new to early modern Europe. The cultures of the ancient Mediterranean had, for example, developed over time a rich and colorful vocabulary with which to make distinctions between different kinds of people and things. The Greeks had famously drawn the line between those who spoke their language and those who did not, dismissing the latter as *barbaroi* or barbarians. The expansion and consolidation of the Roman Empire, meanwhile, with its accompanying far-flung trade networks, had seen the emergence of discourses criticizing "foreign luxuries" and advocating revivals of "aboriginal" Romans' traditional moral virtues. The Roman encyclopedist Pliny, for example, made liberal mention in his gigantic *Natural History* of the origins of natural objects, declaring some "peculiar and vernacular to Italy," while denigrating others as "foreign" and "exotic" luxuries: "so tired do mortals get of things that are their own, and so covetous are they of what belongs to others..."[34] Pliny was, of course, widely read during the humanist revival of the Renaissance, as were countless other Greek and Roman authors. But even more fundamentally, oppositions between things and people seen as belonging "inside" a polity and those seen as somehow "outside" it – exactly the original Greek derivation of "exotic"[35] – seem to have already become commonplaces in many European cultures centuries before the Renaissance, frequently used in a wide range of contexts. It should thus be no surprise that in the mid-fifteenth century, for example, we find the anonymous English author of the lengthy poem *The Libelle of Englyshe Polycye* drawing on these kinds of contrasts to deplore his country's reliance for remedies on the Venetian spice trade: "a man may voyde infirmytee / Wythoute drugges set fro beyond the sea," he insisted.[36] Early modern Europeans thus had a wide-ranging vocabulary of words and concepts about matters near and far to draw on, each with its own array of implications and cultural resonances.

And draw on this vocabulary they certainly did as, over the course of the sixteenth century, commentators came to debate the merits and demerits of those "foreign" natural objects that landed, during this period, on European shores. In the wake of the Columbian Encounter, as existing Mediterranean

---

[34] Pliny, *Natural History*, edited and translated by T. E. Page *et al.* (Cambridge, MA: Harvard University Press, 1938–52), XIV, 200–201; IV, 58–59; and *passim*. See also John Sekora, *Luxury: The Concept in Western Thought* (Baltimore, MD: Johns Hopkins University Press, 1977), 29–38, and Christopher J. Berry, *The Idea of Luxury: A Conceptual and Historical Investigation* (Cambridge: Cambridge University Press, 1994), 45–86.

[35] Henry George Liddell and Robert Scott, eds., *A Greek-English Lexicon*, 9th ed. (Oxford: Clarendon Press, 1951), I, 601.

[36] *The Libelle of Englyshe Polycye: A Poem on the Use of Sea-Power, 1436*, ed. George Warner (Oxford: Clarendon Press, 1926), 19.

trade networks yielded to Iberian competition and new oceanic routes were forged, controversy arose time and time again over imported goods. As has recently been shown, for example, even the traditional spices of the East came under fire, as debates over the shifting structure and control of long-distance trade in the German territories erupted in fierce condemnations of new mercantile companies and of the pepper, cinnamon, cloves, and other spices they conveyed from Asia.[37] Over the course of the century, substances completely new to Europe also began to trickle in from across the globe,[38] and to attract a wide variety of claims and counter-claims concerning their possible uses and effects. One of the earliest of these, guaiac wood from the Andes, was reputed to cure the new "French pox" and thus easily attracted swarms of buyers and sellers, wheelers and dealers. The resulting pamphlet wars between its boosters and its detractors revealed what would be one of the most consistent responses to such new imports: the assertion by physicians that they, and only they, had the necessary qualifications to evaluate the true "nature" of such imports and their possible effects on European bodies.[39] With each new arrival, physicians rushed to pronounce judgment on it, whether urging its prescription in massive doses or condemning its use by Europeans as highly dangerous. The public debate on such tempting new intoxicants and stimulants as tobacco and chocolate, as well as on the numerous other substances that flowed with increasing speed into European ports, thus took on a simultaneously medical and moral tone.[40]

Not all physicians were content to assess the qualities of incoming natural objects on such a case-by-case basis, though. Some began to issue calls for the rejection of all "foreign" or "exotic" substances more generally. Simultaneously, they began to argue for the thorough investigation of their areas'

---

[37] Christine Johnson, "Bringing the World Home: Germany and the Age of Discovery," Ph.D. dissertation, Johns Hopkins University, 2001, 187–247.

[38] The pioneering work of Alfred W. Crosby, Jr. has shown the impacts of biological exchanges, particularly on the Americas and other non-Eurasian areas, both in the Columbian Encounter in particular and throughout world history more generally: see his *The Columbian Exchange: Biological and Cultural Consequences of 1492* (Westport, CT: Greenwood Press, 1972) and his later *Ecological Imperialism: The Biological Expansion of Europe, 900–1900* (Cambridge: Cambridge University Press, 1986). The focus here, however, will be on the flow of species in the opposite direction, namely *into* Europe.

[39] Claude Quétel, *History of Syphilis*, trans. Judith Braddock and Brian Pike (Baltimore, MD: Johns Hopkins University Press, 1990), 29–31.

[40] Wolfgang Schivelbusch, *Tastes of Paradise: A Social History of Spices, Stimulants, and Intoxicants*, trans. David Jacobson (New York: Vintage Books, 1992); Rudi Mathee, "Exotic Substances: The Introduction and Global Spread of Tobacco, Coffee, Cocoa, Tea, and Distilled Liquor, Sixteenth to Eighteenth Centuries," in *Drugs and Narcotics in History*, ed. Roy Porter and Mikuláš Teich (Cambridge: Cambridge University Press, 1995), 24–51. On tobacco and chocolate in particular as two extremely influential New World imports, see Marcy Norton, *Sacred Gifts, Profane Pleasures: A History of Tobacco and Chocolate* (Ithaca, NY: Cornell University Press, forthcoming).

own natural kinds. These apologists for local European nature were not all, like Paracelsus, medical radicals or eccentrics.[41] Nor were they even necessarily among Paracelsus's scattered troop of followers.[42] Rather, they held a wide spectrum of medical opinions, including strict adherence to the Galenic orthodoxy Paracelsus so vehemently opposed. The schoolteacher-turned-herbalist Otto Brunfels, for example, devoted a major portion of the preface to his famous *Herbarum vivae eicones* (Living Images of Herbs, 1530), one of the first great herbals illustrated "from life," to explaining his strong dislike for "alien medicines," and his consequent focus in his tome on "our own herbs."[43] Yet he himself was a thoroughgoing medical humanist, who dedicated much of his career to the translation and editing of works by Galen and Dioscorides, among other classical authors. Likewise, Hieronymus Bock, the author of 1539 *New Kreütter Buch von Underscheydt, Würckung und Namen der Kreütter so in Teutschen Lande wachsen* (New Herbal on the Difference, Effects, and Names of the Herbs that Grow in German Lands), which carefully separated out German plants from foreign (*Welschen*) varieties, was also a careful medical humanist, whose first work had been on Greek and Latin nomenclature.[44] Sixteenth-century compilers of herbals, most (though not all) of whom did indeed strongly favor the rediscovery of local "herbs" over exotic "spices," were a mixed lot. While some, like Brunfels and Bock, were deeply committed medical humanists, others were in fact far more interested in a more prosaic repackaging of existing texts and illustrations for new audiences. Still others drew on already-existing medieval traditions that recognized that often substitutes, or *succedanea* – the term *quid pro quo* was often used – would be needed for unavailable or unaffordable

---

[41] Arber, 255–256, briefly notes their existence, but places them in the context of her discussion on the occult theory of signatures, suggesting her puzzlement with their position.

[42] Andrew Wear, in his thought-provoking article on "The Early Modern Debate about Foreign Drugs: Localism versus Universalism in Medicine," *The Lancet* 354 (July 10, 1999), 150, states that the "argument that local drugs were best was used mainly by the Paracelsian opponents of the establishment medicine based on the teachings of Galen"; however, this may have been the case more in England than in other countries, and more in the sixteenth century than in the seventeenth, by which time numerous quite orthodox physicians had come to use this rhetoric regularly in their local floras. On the English situation, see Ch. 2, "Remedies," in Wear, *Knowledge and Practice*, 46–103.

[43] Otto Brunfels, *Herbarum vivae eicones* (Strasbourg: apud Ioannem Schottum, 1530), 16. He expanded on the topic even further in the preface to the book's subsequent German translation, *Contrafayt Kreütterbüch* (Strasbourg: bey Hans Schotten, 1532), where he devoted an entire section to defending the "usefulness of familiar native (*heymischen*) herbs and medicines." "For what reason," he posed the rhetorical question, "should our herbs not be as good as those from Asia and Africa?" (sig. bii^v).

[44] For some examples of Bock's segregation of German and foreign plants, see Hieronymus Bock, *New Kreütter Buch von Underscheydt, Würckung und Namen der Kreütter so in Teutschen Lande wachsen* (Strasbourg: Wendel Rihel, 1539), sigs. ix^v–x^v, xxi^v, and xxiii^v. For a fuller discussion of Brunfels's & Bock's attitudes towards foreign plants, see Johnson, 223–239.

exotic remedies, especially in the case of the poor, who had long been forced to make use of local herbs simply for reasons of cost.[45] Regardless of their affiliations or intentions, however, many sixteenth-century herbalists seem to have found the language of the revived indigenous-exotic debate compelling, and compilers continued to place a prominent stress on the benefits of local herbs well into the seventeenth century and beyond.

Nor did concerns about the potential negative effects of exotic remedies, and the need to seek out local alternatives for them, remain confined to the German territories, or any other particular area within Europe. This can be seen, for example, in the combative title as well as contents of a 1533 book that renowned French medical humanist Symphorien Champier published in Paris: his *Hortus Gallicus, pro Gallis in Gallia scriptus* (French Garden, Written for the French in France). In this book, Champier declared on his very title page, he would teach "the French to find remedies for all their illnesses in France, not to bring over medicines from foreign sources...."[46] Setting forth a series of rational arguments, many of them plagiarized from Brunfels,[47] as to why medicaments from outside the country were undesirable, he cited various drugs known to be "pernicious and venomous" to Europeans, but perfectly appropriate for those from other regions, explaining the phenomenon by pointing out that since human bodies were influenced by climate, and since climates varied greatly across the world, it made perfect sense that medicines would have different effects on people from different areas.[48] Champier thus clothed his humanist rejection of Arabic medicine in the more genteel garb of a geographical relativism, acknowledging climatic difference and assigning it a key role in medical decision-making. Medieval Western physicians, he argued, had foolishly drifted away from the knowledge of their own environments, adopting foreign medical systems and the drugs they used. Following Brunfels (though, of course, referring to France rather than the German territories), he enjoined his readers to recognize their true identity – "We're in Celtic France, amidst Christians" – and advised his readers that since they were "Christians, not Muslims; French,

---

[45] Siraisi, *Medieval and Early Renaissance Medicine*, 143–147. For an example of learned doctors' attempts to cater to the poor, see Jean Prevost, *Medicina pauperum* (Frankfurt: sumptibus Johannis Beyeri, 1641).

[46] Symphorien Champier, *Hortus gallicus, pro Gallis in Gallia scriptus* (Lyon: in aedibus Melchioris et Gasparis Trechsel fratrum, 1533).

[47] Champier was a notorious plagiarist, although copying passages from other sources had indeed been an accepted medieval practice, from which Champier in the early sixteenth century was still not too chronologically distant. On Champier's tendencies to borrow liberally (and often word-for-word) from others, see Paul Allut, *Étude biographique et bibliographique sur Symphorien Champier* (Lyon: Scheuring, 1859), and Brian Copenhaver, *Symphorien Champier and the Reception of the Occultist Tradition in Renaissance France* (The Hague: Mouton, 1979).

[48] Champier, 4.

not Arabs, or Egyptians, or those born in India, or Palestine," what they really needed to preserve their health were locally-grown medicinals. For Champier, this meant the Mediterranean herbs of the south of France, already part of his French readers' cultural and geographic heritage. Such aromatic herbs as lavender, sage, mint, and thyme were, he opined, the "true spices of Europe."[49] Champier's use, in France, of German rhetoric attacking exotics shows just how flexible appeals to the indigenous-exotic debate could be.[50] Highly malleable, the controversy could be adapted to suit the needs of a wide geographical range of commentators on the natural world, and on its social and cultural implications for the human world.

Over the course of the sixteenth century, then, the distinction between the "indigenous" and the "exotic" became firmly rooted in early modern European discussions about natural objects. In the process, it took on yet further symbolic dimensions. An example may be seen in the herbal of Bartholomäus Carrichter, with the title *Horn des Heyls menschlicher Blödigkeit. Oder, Kreutterbuch* (Horn of Salvation for Human Stupidity. Or, Herbal, 1576).[51] Although this particular work displays a strong Paracelsian influence (Paracelsus himself is thanked in the foreword), it is not atypical of other popular vernacular works, often dealing with astrological themes or the "signatures of things," published in the later sixteenth century. The book's foreword exhibits a concern with many of the themes we have explored: from the politics of the medical profession to a strong interest in the origins of natural objects. More to the point, however, it develops a theme implicit in other such works but not as fully articulated there, namely the religious implications of exploring local nature. Carrichter praised the benevolence of God in establishing a geographical order such that each region would be granted sufficient indigenous medicines to cope with every need. His reliance on the indigenous thus reflected a religious sense of trust: "that God the Lord in this land indeed permits no lack of anything, but rather overflowing plenty of medicines, and that one really would not have any reason to go out of the garden, and to send for medicines in foreign lands...."[52] Furthermore, he

---

[49] Champier, 7, 8.

[50] Ultimately, for example, this rhetoric reached England as well, and was then adapted to the claims of *English* patriotism and geography in turn: see for example Timothie Bright, *A Treatise wherein is declared the Sufficiencie of English Medicines, for cure of all diseases, cured with Medicine* (London: printed by Henrie Middleton for Thomas Man, 1580), discussed in Wear and in Jonathan Gil Harris, *Foreign Bodies and the Body Politic: Discourses of Social Pathology in Early Modern England* (Cambridge: Cambridge University Press, 1998). For another French example, see the discussion of Jean Fernel in James Bono, *The Word of God and the Languages of Man: Interpreting Nature in Early Modern Science and Medicine* (Madison: University of Wisconsin Press, 1995), 89.

[51] Bartholomäus Carrichter, *Horn des Heyls menschlicher Blödigkeit. Oder, Kreutterbuch Darinn die Kreütter des Teutschen lands, ausz dem Liecht der Natur, nach rechter Art der himmelischen Einfliessungen beschriben* (Strasbourg: bey Christian Müller, 1576).

[52] Carrichter, sig. aiiii^r.

argued that anyone who was to scorn the medicines that the Lord dropped "before his door, before the windows and on the ground in front of him" would have to be "a blasphemer."[53] This last remark, with its strong language and implicit accusations of atheism (or worse), suggests a relation between the development of a concern for the local and Reformation natural theology's sympathies for science, in the form of a reverence for the common and familiar things created by God:

...in this book only the herbs of the German land, and of no other country, will be recorded, which is the case for this sole reason, that God the Lord set forth his medicine chests in the entire world, in every single kingdom, principality, region, and parish and therefore for every single man; he has planted them on the mountains, in the valleys, on the plains... behind the fences, and even right before one's door, and thereby built a pharmacy, so that every man on his own land, throughout the year and also every month, could find his own fresh medicines, would not need to buy anything distant, rotten, or spoilt or full of worms, and even less would he need to send at great expense into foreign lands over the mountains or even over the sea for foreign medicines. For the medicines which grow under the stars, beneath which each person himself is born and brought up, are for him (just like bread and meat and drink and everything else that grows up around him) the most customary and useful.[54]

In this vivid vernacular rendering of an argument for the self-sufficiency of the local, we see the emergence of an interest in local nature even more deeply rooted, in its invocation of religion and of the rhythms of everyday life, than that of Paracelsus.

Yet the positing of local nature as an affair of the peasant, as comparable to a concern with "bread and meat and drink and everything else," reveals some of the aspects of local nature study that were to seem less attractive, even potentially threatening, to sixteenth- and seventeenth-century physicians. In the work of many sixteenth-century herbalists, the indigenous came to represent the domain of the peasant, not just the scholar; indeed, scholars were unnecessary in the worldview Carrichter presented, since each individual could learn to identify herbs on his own, just as God had presented them to him. Numerous other examples tell the same story: though a strong interest in the "native" could be found in both popular and learned writings about nature, it remained controversial and potentially divisive. It is thus that, when medical professors began to compile the earliest local floras in the early seventeenth-century university, they would find it necessary to defend their very attention to the indigenous, to the "common" and "familiar." A tone of uneasiness is marked in many of the early local floras, as their

[53] Carrichter, sig. aiiii^r.
[54] Carrichter, sig. aiii^v. On religious language in vernacular natural-historical works, see Kathleen Crowther-Heyck, "Marvellous Secrets of Nature: Natural Knowledge and Religious Piety in Reformation Germany," *Isis* 94 (2003): 253–273.

authors argued that studying the local was in no way incompatible with learned medicine.

In short, by the eighteenth century, the quarrel between the "indigenous" and the "exotic" had become a commonplace within early modern European culture, perhaps in some ways even as familiar a theme as that of the battle between the ancients and the moderns.[55] When books appeared on "exotic" themes, for example, their authors often found it necessary to allude to the controversy. One author even indexed the topic.[56] The English vicar Robert Burton included a lengthy excursus on it in his famous *Anatomy of Melancholy*.[57] It reappeared quite frequently as a dissertation topic for medical students.[58] Writers of numerous treatises on the fashionable new beverages of coffee, tea, and chocolate referred to itrepeatedly.[59] Even Voltaire, the doyen of the French Enlightenment, turned it to his purposes in his entry on "China" in his notoriously satirical *Philosophical Dictionary*: "We go to China for china-clay as if we had none of our own; for fabrics as if we lack fabrics; for a little herb to absorb water as if we had no simples in our climes."[60] Voltaire, of course, had relatively little interest in medicine or botany per se. Poking fun at tea, though – "a little herb to absorb water" – offered him an enjoyable and witty way to criticize his contemporaries' predilection for the exotic. The debate between the exotic and the indigenous had become a

[55] For several reassessments of this classic debate, which preoccupied scholars for centuries, see Joseph Levine, "Ancients and Moderns Reconsidered," *Eighteenth-Century Studies* 15 (1981): 72–89, and Joan DeJean, *Ancients Against Moderns: Culture Wars and the Making of a Fin de Siècle* (Chicago, IL: University of Chicago Press, 1998).

[56] See for example Michael Bernhard Valentini, *Polychresta exotica* (Frankfurt: sumptibus Johannis Davidis Zunneri, 1700), a collection of various dissertations on exotic substances of all sorts. In the "Index Rerum & Verborum" at the back of the book, he specifically called attention to his discussion of the controversy in his preface, including entries on "Exotica an rejicienda?" ("Should exotic things be rejected?") and "Domestica remedia an sufficiant?" ("Are domestic remedies sufficient?").

[57] Robert Burton, *The Anatomy of Melancholy* (Oxford: Printed by John Lichfield and James Short, 1621), 430–437.

[58] See for example Olaus Borrichius, *De usu plantarum indigenarum in medicina* (Copenhagen: literis & impensis Joh. Phil. Bockenhoffer, 1688); J. M. Hengstmann, *Dissertatio medica inauguralis de medicamentis Germaniae indigenis sufficientibus* (Helmstadt: litteris Pauli Dieterici Schnorii, 1730); Benjamin Gottlieb Albrecht, *Dissertatio inauguralis medica de aromatum exoticorum et nostratium praestantia* (Erfurt: typis Heringii, 1740). These may always, of course, have been written by the professors presiding over the dissertation defense, as was common at the time; see Gertrud Schubart-Fikentscher, *Untersuchungen zur Autorschaft von Dissertationen im Zeitalter der Aufklärung* (Berlin: Akademie-Verlag, 1970).

[59] See Piero Camporesi, *Exotic Brew: The Art of Living in the Age of Enlightenment*, trans. Christopher Woodall (Cambridge: Polity Press, 1994), and Jordan Goodman, "Excitantia: Or, How Enlightenment Europe Took to Soft Drugs," in *Consuming Habits: Drugs in History and Anthropology*, eds. Jordan Goodman, Paul E. Lovejoy, and Andrew Sherratt (London: Routledge, 1995), 126–147.

[60] Voltaire, *Philosophical Dictionary*, edited and translated by Theodore Besterman (Harmondsworth, UK: Penguin, 1971), 112.

familiar enough trope that even those far removed from active engagement in natural inquiry could deploy it almost effortlessly.

It is worthwhile, though, to look closely at yet one more example of the use of this rhetoric. The compiler of one abridged book of foreign travels, a Dutchman by the name of Olfert Dapper, employed coarse metaphor to evoke the debate, much as Carrichter had done a century and a half earlier. In one passage, he made fun of anyone who, he claimed, was "merely content with garlic and onions, the kind that grow before his own door, and does not look around, to see whether there are also people living on the other side of the mountain, who enjoy cinnamon and sugar...."[61] This usage of the language of contrast between local and exotic products shows not only the persistence, and compelling interest, of the metaphor into the seventeenth and eighteenth centuries, but also its versatility. Here this author, a popularizer of the exotic and compiler of numerous works on foreign voyages, employed the vernacular not to defend the use of things "that grow before his own door," but rather to reject them as insufficiently cosmopolitan. Here garlic and onions appear as emblematic of the "low" status of local natural history, of its connections to peasant worlds.[62] Spurred by "curiosity" and "fiery desire", the traveler in Dapper's vision leaves his native land and happily travels throughout the world, collecting everything about "cities, animals, herbs, trees, minerals and those kinds of things."[63] Yet even Dapper's celebration of travelling, and of exotic nature, displays more than a hint of defensiveness, since he notes the existence of, and feels compelled to respond to "those who really want to throw out the baby with the bath water, i.e. those who reproach all travels...."[64] The persistence of these kinds of anxieties, and this kind of rhetoric, must be seen as helping to pave the way for the outbreak, later in the eighteenth century, of the famous "Dispute of the New World," which saw the famous French naturalist Buffon quarreling

---

[61] Olfert Dapper, *Exoticus Curiosus* (Frankfurt & Leipzig: bey Michael Rohrlachs seel. Wittib und Erben in Liegnitz, 1717), sig.)(3r –)(3v. Though Dapper himself was Dutch, his works, with their evocations of wonders abroad, were translated into numerous languages and were thus in fact quite typical of the kinds of books available to audiences throughout Europe by this point, in no small part owing to the role played by the Amsterdam printing presses. On early eighteenth-century Dutch exoticism, see Benjamin Schmidt, "Inventing Exoticism: The Project of Dutch Geography and the Marketing of the World, circa 1700," in *Merchants and Marvels: Commerce, Science, and Art in Early Modern Europe*, ed. Pamela H. Smith and Paula Findlen (New York: Routledge, 2002), 347–369.

[62] Allen J. Grieco, "The Social Politics of Pre-Linnaean Botanical Classification," *I Tatti Studies: Essays in the Renaissance* 4 (1991): 131–132, 135, 140.

[63] Dapper, sig.)(4v -)(5r. On curiosity and travel, see Neil Kenny, *Curiosity in Early Modern Europe: Word Histories* (Wiesbaden: Harrassowitz, 1998) and Justin Stagl, *A History of Curiosity: The Theory of Travel, 1550–1800* (Chur, Switzerland: Harwood Academic Publishers, 1995).

[64] Dapper, sig.)(3r.

with American founding father Thomas Jefferson over the size and putative "degeneracy" of New World species versus those of the Old.[65] By that point, the polemics we have traced had become so much a part of both learned and popular discourse on natural variety that their long trajectory had effectively become invisible.

Over the course of the sixteenth and seventeenth centuries, then, the debate over the "indigenous" versus the "exotic" became firmly entrenched in European culture in general, and in disputes over the natural world in particular. While some who drew on the debate used it to promote the reach of foreign trade throughout the world, with the wealth of potentially useful commodities that trade made available, others reacted severely against the perceived influx of "exotic" substances in Europe and called instead for the reassertion of local resources. What might have remained a simple rhetorical dualism became instead a matter of serious practical import, and the topic of sharp controversy, in which a wide range of individuals were summoned to take sides. In this charged atmosphere, the study of natural objects took on particular tensions. As early modern natural inquirers, then, sought to define their own objects of study, they found themselves caught in the midst of a complex and shifting set of concerns over natural origins.

"INDIGENOUS MEDICINE"

Beginning in the late sixteenth century and proceeding into the seventeenth, several authors went so far as to compose lengthy treatises dedicated solely to the mission of defending their countries' natural worlds – and, in the process, the European "indigenous" itself – as a serious topic of discourse. Whereas most other chroniclers of local nature had entrusted at most a few prefatory pages to the topic of the indigenous–exotic debate, each of these new authors devoted the full contents of an entire book to the claims of the "indigenous" and the systematic rebuttal of the exotic's counterclaims. Though such works came to appear in a number of northern European locations, two that were published over the course of several decades in the seventeenth-century Netherlands – Jan van Beverwyck's *Autarkeia Bataviae, sive introductio ad medicinam indigenam* (Batavian Autarky, or, an Introduction to Indigenous Medicine, 1644), and Lambert Bidloo's *Dissertatio de re herbaria* (Dissertation on Botanical Matters, 1683) – are especially revealing of these efforts to establish a full scholarly viability for the European "indigenous." With their self-conscious theorizing about nature and the native, these two treatises offer the modern reader particular insight into

---

[65] Antonello Gerbi, *The Dispute of the New World: The History of a Polemic, 1750–1900*, trans. Jeremy Moyle (Pittsburgh, PA: University of Pittsburgh Press, 1973).

how, and why, "indigenous" nature came to be viewed as a topic worthy of study in its own right.

In 1644, Jan van Beverwyck, a physician and town councillor in the Dutch port of Dordrecht, published his *Autarkeia Bataviae*.[66] The title of this book deserves a word of explanation. Like other Dutch intellectuals of his day, van Beverwyck was fascinated by tales of the ancient Batavi, the Germanic tribe said to have originally inhabited the region of the Low Countries before the arrival of the Romans. Earlier humanist writers, in the throes of the Dutch Revolt against the Spanish and Austrian Hapsburgs, had seized on the Batavi, who were reported to have fiercely resisted the Romans, as symbols of Dutch national pride and hoped-for independence.[67] By the mid-seventeenth-century it became standard scholarly practice to use the term "Batavian" as, for all practical purposes, synonymous with "Dutch." The label's earlier patriotic resonances, however, continued to echo for readers. With the use of the term thus came a sense not only of Dutch distinctiveness – by the middle of the seventeenth century, with the merchant republic's unparalleled success in international commerce, that could no longer be in doubt – but of the autochthonous nature of the region's inhabitants, who had lived on the land before the Romans' (or the Hapsburgs') arrival. The term thus summoned up strong images of prior presence, together with resistance to empire.[68]

Van Beverwyck's use of the term "autarky" was likewise freighted with meaning. What the word literally meant, in its original ancient Greek usage, was a situation of utter economic self-sufficiency, usually encountered only in wartime, in which trade with other nations had ceased completely. Central European cameralists, like those of the various tiny German territories, cut off from direct access to the new colonial trades, would come to embrace this concept over the next century, turning necessity into a virtue.[69] Hence

[66] Johan van Beverwyck, *Autarkeia Bataviae, sive introductio ad medicinam indigenam* (Leiden: ap. Joh. Maire, 1644). This book was actually a reworking of a book he had published two years earlier in Dutch, his *Inleydinge tot de hollandse geneesmiddelen* (Dordrecht: voor Jasper Gorissz., 1642); however, the Latin version became – for obvious reasons of linguistic accessibility to an audience outside the United Provinces – more popular across Europe. It was cited by numerous compilers of local floras, especially German and French ones, and continued to be cited well into the eighteenth century. On van Beverwyck's career, see Christian Wilhelm Kestner, *Medicinisches Gelehrten-Lexicon* (Jena: bey Johann Meyers seel. Erben, 1740), 110.

[67] See Simon Schama, *The Embarrassment of Riches: An Interpretation of Dutch Culture in the Golden Age* (Berkeley: University of California Press, 1988), 75–80, and I. Schöffer, "The Batavian Myth during the Sixteenth and Seventeenth Centuries," in *Britain and the Netherlands, V*, eds. J. S. Bromley and E. H. Kossmann (The Hague: Nijhoff, 1975), 78–101.

[68] Benjamin Schmidt, *Innocence Abroad: The Dutch Imagination and the New World, 1570–1670* (Cambridge: Cambridge University Press, 2001), 74–76.

[69] Edgar Schorer, "Der Autarkiebegriff im Wandel der Zeiten," *Jahrbuch für Gesetzgebung, Verwaltung und Volkswirtschaft im Deutschen Reich* 65 (1941): 47–82, and D. C. Coleman, ed., *Revisions in Mercantilism* (London: Methuen, 1969). Concepts of autarky have generally

van Beverwyck's use of the term, given his residence in a country that had recently come to profit so greatly from international commerce, is indeed striking. What van Beverwyck seems to have intended to evoke through his book's title was a sense of the patriotic necessity, which he urged upon his countrymen, to rush to the defense of that "indigenous medicine" he recommended in his subtitle. The Dutch, as he saw it, had been overwhelmed over the previous decades with "exotic" influences; the only way to restore balance and harmony would be to return to an earlier presumed state of self-sufficiency, in which the Dutch would again relearn to rely on their own "indigenous" or "domestic" resources.

Van Beverwyck's approach to the debate between the "indigenous" and the "exotic" drew heavily on humanist ideals and methods. He cited one ancient author after another to support his claims that the perils of "foreign" substances had been recognized even in Greek and Roman times. He used the materials and environments of the ancient Mediterranean world as a primary point of reference, comparing and contrasting these with their northern European counterparts. And, he felt, the ancient Batavi would have agreed heartily with the classical authors whose strictures against exotics he cited. "Indeed I for my part could not believe that the ancient Batavians," before entering into world trade "so that they might return burdened with the spoils of the Orient, would not have made use for preserving health, or recovering it, of their indigenous herbs." Indeed, he observed, "they would otherwise have been more stupid than . . . cats and dogs, who know about domestic remedies, and do not set sail in search of grass or mint."[70] Van Beverwyck traced the history of modern commerce back to the Venetians, who bought their "exotics" from Egypt, subsequently yielding their primacy in trade to the Spanish, and ultimately to the Dutch themselves.[71] Putting the controversy over exotics into historical perspective, Beverwyck thus felt free to cite a wide range of authors, from the ancients to contemporaneous raconteurs of Dutch voyages to both the East and West Indies, to support his case for repudiating "exotic" in favor of "indigenous" or "domestic" nature.

Why did van Beverwyck so strongly condemn the use of exotics? He gave numerous reasons. God would, he felt, "never have forced miserable mortals to fetch things from distant lands," "lands warmed by another sun."[72] Exotics had only become popular in his day because Europeans, curious and gullible, had let themselves be deceived by the glamor attached to exotic imports, mistaking high prices for true value. Referring to "exotic medicaments" in particular, van Beverwyck sourly followed Pliny in observing that they were "more helpful for enriching pharmacists, than for curing sick

flourished during times of war, and Europe was at this point still in the throes of the Thirty Years' War; as we shall see, however, the way in which van Beverwyck developed this concept went beyond any wartime setting.

[70] van Beverwyck, 53–4.   [71] van Beverwyck, 54.   [72] van Beverwyck, 39–40.

people."[73] People who bought exotic remedies such as balsam or tamarind were, he felt, all too commonly sold a bill of goods; the herbs and roots they purchased were often adulterated with other substances, or had simply gone stale from too much transit time. As a result, not only were these exotic substances unnecessary, but they could, even worse, be positively damaging to health.[74]

In contrast to exotics, van Beverwyck argued, "indigenous" natural objects were safer, more reliable, and generally superior for all purposes. "On the contrary [i.e. to exotics], nothing can be more certain than indigenous plants, which we see every day."[75] Like Paracelsus, van Beverwyck seems to have seen living beings as existing in a close relationship or "sympathy" with their surrounding natural environments, which affected them in ways more deeply than human beings could ever hope to understand. Each region, in particular, had its own specific endemic diseases, which only native medicines could cure. European plants and animals, insisted van Beverwyck, thus shared a special bond with European people, "since they live under the same sky with us, and in the same soil, and they consume the same food, known to us, and they assume a nature harmonious to our nature."[76] This harmonious relationship ensured, for example, that a food, drink, or medicine could be consumed and would not injure, or be violently rejected by, the body of the person or creature who ate it. More generally, this material harmony of influences and ingredients spoke to a deeper sense of natural and divine order, in which living beings and indeed nonliving objects as well "fit" their environments in a perfect match.

Travel, whether of men, beasts, or plants, was seen as disrupting this harmony. Van Beverwyck cited the well-known fact that tropical plants brought to Europe, like aloes, tended to do badly in their new surroundings. He attributed this not only to harsh European winters, but to a more fundamental imbalance. Transplanted species "fight with a hostile sky and soil," he explained, "and they're not able to enjoy their native and familiar food, and thus cheated out of their spirit, they gradually wither and eventually die."[77] As with plants, so with people; Europeans too, van Beverwyck felt, tended to degenerate in their morals, customs, and general health when they travelled outside their native lands.[78]

It is in this context that van Beverwyck proposed a renewed attention to the "indigenous" natural products of the Netherlands. Holland, he maintained, was a virtual "storehouse of fertility," blessed with "affluence" in its natural

---

[73] van Beverwyck, 68. On the keen interest of Dutch physicians in natural history, see Harold J. Cook, "The Cutting Edge of a Revolution? Medicine and Natural History near the Shores of the North Sea," in *Renaissance and Revolution: Humanists, Scholars, Craftsmen, and Natural Philosophers in Early Modern Europe*, ed. J. V. Field and Frank A. J. L. James (Cambridge: Cambridge University Press, 1993), 45–61.

[74] van Beverwyck, 74–75, 72, 71, 91.   [75] van Beverwyck, 71.   [76] van Beverwyck, 76–77.

[77] van Beverwyck, 105–108, 43.

[78] van Beverwyck, 112; see also Schmidt, *Innocence Abroad*, 281–310.

endowments as well as its banking institutions.[79] Van Beverwyck admitted that Holland did, indeed, owing to its small size and geographical situation, conspicuously lack some of the natural resources enjoyed by neighboring countries, such as metal deposits to be mined, or adequate forest cover to burn for fuel; but he argued that if studied closely enough, the land would reveal sufficient "indigenous" resources to cover all of its needs. Take, for example, the case of fuel; even in the absence of sufficient quantities of wood, peat deposits amply sufficed to meet Dutch energy needs.[80] He argued against the excessive importation of colonial sugar, devoting an entire section of the book to the advantages of native honey as a substitute.[81] Similarly, he proposed that those tempted by exotic oils simply use butter, the product of the thriving Dutch dairy industry; he did acknowledge butter's tendency to go rancid, but discussed possible preservatives.[82] He scorned the new foreign drinks, lauding Dutch beer instead.[83] And he pointed to Dutch herbs which were known to cure local diseases, arguing that if only physicians and other patriotic individuals turned their minds to the task, borrowing a leaf or two from wise rustics in the process, the Netherlands could be shown to possess a full complement of "indigenous" resources in this regard as well.

In short, van Beverwyck drew on the themes of the indigenous–exotic debate, drawing out the moral and social implications of "foreignness" for both plants and people, to articulate a strong defense of "indigenous" European nature. Though his focus was primarily on Dutch examples, the book proved much more widely influential; over the course of the next century, it was frequently cited by authors from both France and the German territories. In particular, naturalists cited the book as support for the compiling of local floras or "catalogues of indigenous plants" from a number of different European regions. Clearly, what they took from the book was not so much any conviction of the indispensability of Dutch nature in particular, as rather the broader point that van Beverwyck was making: namely, that the indigenous–exotic debate had raised crucial intellectual problems, which Europeans could best address by making a thorough study of their own "indigenous" natural productions.

[79] van Beverwyck, 5.   [80] van Beverwyck, 30, 18–19.
[81] van Beverwyck, 99–103. Danish physician Thomas Bartholin similarly proposed honey as a substitute for sugar in his *De medicina Danorum domestica dissertationes X* (Copenhagen: typis Matthiae Godicchenii, 1666), a work similar to that of van Beverwyck in many ways; see Martha Baldwin, "Danish Medicines for the Danes and the Defense of Indigenous Medicines," in *Reading the Book of Nature: The Other Side of the Scientific Revolution*, eds. Allen G. Debus and Michael T. Walton (Kirksville, MO: Sixteenth Century Journal Publications, 1998), 169. Bartholin seems to have understood the phrase "domestic medicine" not as referring solely to the use of household remedies (gendered female), but also of *local* ones as well, by physicians in addition to laypeople; for a contrast with a later period, see Charles E. Rosenberg, "Medical Text and Social Context: Explaining William Buchan's *Domestic Medicine*," in *Explaining Epidemics and Other Studies in the History of Medicine* (Cambridge: Cambridge University Press, 1992), 32–56.
[82] van Beverwyck, 103–105.   [83] van Beverwyck, 25–27.

In 1683, several decades after the appearance of van Beverwyck's book, a new treatise on the topic was published. Its title – "Dissertation on Botanical Matters" – told very little about its contents.[84] But the way in which it was presented to its readership provides us with some clues as to why, despite its dry title, this treatise entered directly into the indigenous–exotic debates. For the treatise was bound together with, effectively as an (extremely extended) preface and introduction to, the first explicitly local flora of the Netherlands: the famous botanist Jan Commelin's *Catalogus plantarum indigenarum Hollandiae* (Catalogue of the Indigenous Plants of Holland).[85] Commelin, at first glance, might seem far from an obvious candidate for the authorship of such a book. A merchant and importer of exotic medicines by profession, who had done well enough for himself to be appointed to various posts in the Amsterdam city government, he had profited greatly from exactly the kind of enthusiasm for exotics van Beverwyck had so decried.[86] And in the same year his *Catalogus* was published, Commelin had just been selected as director of the new Amsterdam botanical garden, which would eventually under his leadership come to possess one of the widest selections of exotic species of any garden in Europe.[87] But for him as for van Beverwyck, his familiarity with exotics and with the indigenous–exotic debate had clearly only whetted his curiosity about "indigenous" nature. Basing his *Catalogus* on botanizing he'd done around his own country estate in Haarlem, Commelin recruited the Amsterdam apothecary Lambert Bidloo to introduce it to his readers. Far from supplying a brief and merely ornamental preface, though, Bidloo ended up contributing a full-fledged treatise, one that would engage many of the issues surrounding his own and Commelin's careers.

Bidloo's treatise directly addressed itself to the readers of Commelin's indigenous plant catalogue. "If you, dear reader, look at Commelin's... volume, you will see a tiny book, but given the amount of labor assembled for it, a work quite large enough. For [to produce it] indeed what a number of fields, forests, thickets, hills, and beaches had to be crawled through!"[88]

[84] Lambert Bidloo, *Dissertatio de re herbaria* (Amsterdam: apud H. & viduam T. Boom, 1683).

[85] Jan Commelin, *Catalogus plantarum indigenarum Hollandiae* (Amsterdam: apud H. & viduam T. Boom, 1683).

[86] F. W. T. Hunger, "Jan of Johannes Commelin," *Nederlandsch Kruidkundig Archief* (1924), 187–202. Commelin's fascination with plants from much warmer locales, for example, is evidenced in his *Nederlantze Hesperides, Dat is, Oeffening en Gebruik Van de Limoen- en Oranje-Boomen; Gestelt na den Aardt, en Climaat der Nederlanden* (Amsterdam: by Marcus Doornik, 1676), which aimed to teach the Dutch how to plant and acclimatize citrus trees.

[87] D. O. Wijnands, *The Botany of the Commelins* (Amsterdam: A. A. Balkema, 1983).

[88] Bidloo, 3. On the trope of natural knowledge as a secret that has to be laboriously hunted after, see William Eamon, *Science and the Secrets of Nature: Books of Secrets in Medieval and Early Modern Culture* (Princeton, NJ: Princeton University Press, 1994), 269–300.

Praising Commelin's achievements in scrutinizing the "corners, valleys, and remote vaults" of their native land, and thereby finding "many indigenous plants hitherto unknown," the treatise moved on to explore a number of contemporary controversies in botany: most noteworthy among these, of course, the controversy over Commelin's chosen object of study in his catalogue, "indigenous plants." Though Bidloo occasionally left this topic to explore other botanical trains of thought, he always circled back to the indigenous–exotic debate; and it is worth seeing what he had to say.

Bidloo's stance on this topic was, on the whole, quite similar to van Beverwyck's. Indigenous species, he felt, had been unjustly ignored in the rush to study and consume all things exotic. He attributed the popularity of exotic substances to a craving for novelty ("for one kind of person, nothing will suffice unless it's new"), and compared changing tastes in food, drink, and medicines to those in the fashion world, referring contemptuously to girls' dresses as an example of this, adding the contemptuous remark "away with you Dutch herbs! family doctors are now prescribing tea, coffee, and chocolate."[89] Bidloo warned of excessive passion for exotics, ominously hinting, like van Beverwyck had, that this trend heralded decline: "due to the wares of foreigners, weakness, luxury, and gluttony are now stealing over our people, as happened to the Romans in their day...."[90] To illustrate his point, he cited the case of tobacco: "Have men increased their longevity in our age, in which the use of nicotine has increased so greatly? On the contrary, as seen from examination of cadavers of the dead, as many anatomists have noted."[91]

Bidloo's analysis of the roots of the problem likewise mirrored van Beverwyck's. Objects and environments, he insisted, were linked in an intricate balance, which must not be disrupted. "The soil and the sky of every region mutually harmonize together and are connected, for men as for plants, in a universal relation on all sides."[92] Consuming foreign substances incurred the great risk of violating this natural order. Bidloo reported, for example, that plants from the Indies, if eaten by Europeans, commonly caused bloody diarrhea, vomiting, paralysis, "and other serious symptoms." For Bidloo, the same general rule held true in Europe, just to a lesser degree, since distances were shorter and environmental differences therefore less extreme; thus an Englishman would probably get sick on a diet of Norwegian fish.[93] Bidloo acknowledged that proponents of exotic medicines had begun to call these kinds of arguments based on affinities and "sympathies" between

---

[89] Bidloo, 34.
[90] Bidloo, 24. On specifically Dutch concerns over luxury and excess (*overvloed*), see Schama, *passim*; on early modern worries about luxury more generally, see Sekora; Berry; and Maxine Berg and Helen Clifford, eds., *Consumers and Luxury: Consumer Culture in Europe 1650– 1850* (Manchester: Manchester University Press, 1999).
[91] Bidloo, 36.   [92] Bidloo, 8.   [93] Bidloo, 8–9.

"earth, water, and sky" in doubt, questioning them both as to their rational grounds and state of empirical proof. Admitting that indeed he could not "prove" the connections he saw with any kind of "mathematical" certainty, he nonetheless maintained that the overwhelming weight of the evidence, and of common sense itself, was on his side.[94] Here too, then, Bidloo chose to accept van Beverwyck's basic theoretical model, arguing that it was the only one that made sense of the observations Europeans had accumulated about the historical interactions between objects and their environments.

If one examines Bidloo's "dissertation" closely, though, signs can be seen that distinctions between "indigenous" and "exotic" were becoming increasingly difficult to uphold, for those involved in the serious pursuit of natural history or indeed for anyone else who had seriously thought the issue through. Bidloo observed, for example, that though many exotic plants grew only feebly if at all upon transplantation to the Netherlands, a few had in fact, after solicitous care from their gardeners, eventually succeeded in acclimatizing to their new environment, where they were now thriving quite nicely. "Many things from lands and skies quite unlike our climate are now growing here abundantly, as if in their own natural soil. Aren't the Canadian chrysanthemum and the Peruvian potato ... now grown in our fields?" In the case of the potato, what had once been a strange import had now become a staple, been given its own Dutch name (*Aard-Appel*) as if it had always been there, and become fully naturalized into Dutch life.[95] Bidloo reported that he could name at least 600 other such cases; however, he did not do so, but contented himself with referring to the notorious example of tobacco, of whose hazards he had earlier warned. As he pointed out, entrepreneurial Dutch farmers had begun to cultivate tobacco plants with surprising success. "What about *Nicotianum*, occupying vast fields of ours, and very happily springing forth?"[96] Nor were commercial crops the only neophytes to prosper; as Linnaeus would shortly thereafter remark, the introduced medicinal plant *Acorus calamus* now grew wild and "luxuriant along the Dutch canals."[97] If foreign species could clearly not only find acceptance among Dutch people, but also thrive in Dutch soils, what did this say about the relationship between the indigenous and the exotic?

---

[94] Bidloo, 9.

[95] Bidloo, 73; cf. Redcliffe Salaman, *The History and Social Influence of the Potato*, revised ed. (Cambridge: Cambridge University Press, 1949).

[96] Bidloo, 73. On the origins of Dutch tobacco cultivation during this period, see H. K. Roessingh, "Tobacco Growing in Holland in the Seventeenth and Eighteenth Centuries: A Case Study of the Innovative Spirit of Dutch Peasants," *Acta Historiae Neerlandicae* 11 (1978): 18–54.

[97] Cited in K. V. Sykora, "History of the Impact of Man on the Distribution of Plant Species," in *Biological Invasions in Europe and the Mediterranean Basin*, ed. F. di Castri, A. J. Hansen, and M. Debussche (Dordrecht: Kluwer, 1990), 46.

These kinds of concerns can be seen as coming to the fore in the very way that Bidloo chose to define the "indigenous." Whereas van Beverwyck had never fully stipulated what he meant by the term, establishing its parameters more through example and through stark contrast with the "exotic" than by explicit definition, Bidloo seems to have felt compelled to clarify how he understood the term. He did so quite early in the book, on its second page. By the term "indigenous," he commented, he understood "not only these things, which originated here of their own accord since before the memory of men, but also those which, cast down here from other shores, owing to their frequent cultivation here, having grown accustomed to our sky and soil, have now been granted citizenship...."[98] By explicitly including acclimatized exotics in this definition, to justify their inclusion in Commelin's catalog, Bidloo thus framed a generously wide understanding of the scope of the "indigenous." In the process, he highlighted the increasing difficulty of distinguishing between natural objects based on their geographical origin, in a world where species had come to be interchanged on an ever-more-frequent basis. "Many *exotica* are *indigena* by cultivation.... Indeed it would be a tough and unpropitious business without doubt, to determine which plants grow here and not elsewhere, whether of their own accord, or by seeds that have been brought here...."[99] And indeed Commelin did go on in his inventory of Dutch plants to list not just *Acorus* by the canals, but tobacco itself, "lots of it, in the fields by Amersfoort."[100] For Bidloo and Commelin, even though they made plentiful use of the indigenous–exotic debate as a way of justifying their efforts, distinctions between the "indigenous" and the "exotic" could not, in all honesty, actually be drawn so clearly. As they acknowledged, the categories were permeable, and travel between continents could, and did, change them.

As the case of Bidloo and Commelin thus shows, the desire to study the "local" and the "indigenous" natural phenomena of early modern Europe was thus by no means a self-evident process, but was rather embedded in a process of debate within early modern Europe. This debate, which came to be framed through the polarities of the "indigenous" and the "exotic", but ultimately challenged them, kept coming to the fore again and again in early modern Europe. As rumors and reports filtered in, from far-off parts of the globe, of different people and creatures elsewhere, and as new material objects began to substantiate some of these rumors, Europeans struggled to make sense of the "exotic" phenomena they encountered. And while some

---

[98] Bidloo, 4. In recent years, a vast literature has sprung up on the concept of citizenship; however, much of it, in the early modern period at least, is devoted solely to the analysis of individual political theorists, rather than actual practices. See for example Derek Heater, *A Brief History of Citizenship* (New York: New York University Press, 2004), especially 50–64.

[99] Bidloo, 5.   [100] Commelin, 2, 78.

reacted favorably to these phenomena, others seem to have reacted *against* the exotic in and of itself. In their reactions against the exotic, some went further, going so far as to elevate "indigenous" European natural objects to an importance they had not previously possessed. In Amsterdam, Jan Commelin compiled the first thorough inventory of the flora of Holland. And in London, Nicholas Culpeper, protesting against the "outlandish," launched his spectacularly successful herbal. The reevaluation of the European "indigenous" had truly begun.

2

Field and Garden: The Making of the Local Flora

One morning in 1727, a procession assembled in Altdorf, a small German town. The *Rector Magnificus* of the university was there, as were assorted deans, "all wrapped in their new and splendid robes," professors, doctors, masters, and many other "citizens of the Academy." They marched from the Theologicum, the lecture-hall of the theologians, over to the Welserianum, an auditorium newly decked-out for the occasion. There, to the accompaniment of tubas and tympanies, they listened to a chanted ode and to an "Oration on the Origin, Progress and Destiny of the Medical Garden of the Altdorf Academy." Then, after mutual congratulations, they all went home. This was not the end of the day's festivities, though. Around noon, a more select group of professors met in the botanical garden's greenhouse. Here, according to Johann Jakob Baier (the director of the Altdorf *hortus medicus* at the time and organizer of the festivities), they "did not scorn to be made partners in botanical gaiety," but engaged their spirits in "licit joy" through music and conversation, "peacefully" (so Baier assured his readers) until late into the night (see Figure 3).[1]

The botanical garden of Altdorf, founded in 1626, had just celebrated its centennial (albeit a year late). And much had indeed happened during the past century. Over the course of the intervening period, the small walled town of Altdorf – about 20 km away from the thriving trading center of Nuremberg – had grown into a focal point for the new sciences. Simultaneously, tiny Altdorf's plant world had come to be one of the most highly studied – and written about – in all of Europe, indeed in the entire world at the time. A hundred years had produced a lasting tradition devoted to the compilation and publication of what would later come to be called "local floras." To create these documents, the countryside surrounding the town had been canvassed again and again for its diverse plant species. Altdorf was not the only early modern European town that had come to enjoy this curious privilege of having its local plants scrutinized in detail and recorded for posterity. During this period, many other municipalities (at first largely in the scattered territories of the Holy Roman Empire, then increasingly elsewhere

---

[1] Johann Jakob Baier, *Horti medici Acad. Altorf. historia curiose conquisita* (Altdorf: typis Iod. Guil. Kohlesii, 1727), sig.)(5v and)(6r.

Figure 3. Academic Procession through Altdorf. Engraving by Johann Georg Puschner, 1718. The line of faculty and students, marching in order of rank, weaves its way through the *Marktplatz* of the small town, past the church on the left. Courtesy of the Stadtarchiv Altdorf.

in Europe and ultimately the colonies) had had local floras discussing their "indigenous" plants written. What united the areas that came to acquire such local floras, and separated them from those that didn't, was their possession of several key institutions within Baroque Europe: first the early modern town itself; second the Latin-speaking enclave of the university or academy; and third the recently-developed botanical garden, with its tantalizing array of exotic species from newly-trafficked continents. It was in these contexts that the "local flora," over the course of the seventeenth century, came to be standardized and institutionalized, and ultimately exported to the rest of the world. By the eighteenth century, some ambitious European voyagers to far-away sites had even begun to write fully-fledged local floras to take inventory of the plant species they found. The local flora had gone global.

So what, in fact, *was* a "local flora"? The document we now call by this name was in its most basic form a list or inventory of the plants to be found growing within a given area, usually defined in early modern exemplars as that within a radius of three, four, or five miles around a town. Local floras were generally very small and modest documents, usually about the size of the palm of one's hand, and rarely numbering more than a hundred pages. They presented information about the plants of a given area in a highly concise form, often cryptically abbreviated. A typical entry, for example, might consist of a plant's Latin name, followed by several variant learned names for it, also in Latin, its popular or vernacular name, and finally a brief notation of the kinds of areas where the plant could be found. For example, plants might be listed as located within the walls around an early modern town; in the farmlands that typically surrounded it; or in nearby swamps or forests, by roadsides or on mountaintops. This is the format that became established in the early modern German territories, where many of the earliest local floras were written, and that subsequently came to form a model for plant catalogues across the globe.

There is little agreement as to when or where the first local floras appeared. Those who have commented on this issue, chiefly botanists, have tended to assign their choices along national lines. For example, Anglo-American authors have tended to point to John Ray's catalogue of plants in the Cambridge countryside (1660) as the first "true" local flora,[2] whereas Central European writers have placed the onset of the genre much earlier, citing examples from the German territories of the late sixteenth and early seventeenth centuries.[3] Contributing to the confusion, a variety of different criteria have been used to determine what counts as a truly "local flora." One crucial indicator, for example, has been the percentage of species listed in a flora that modern botanists consider "indigenous" to an area (i.e. naturally growing in the area, as opposed to garden or other cultivated plants introduced through human intervention). Even more sharply restricted is the group of species considered "endemic" to an area (i.e. not only naturally growing in that area, but to be found nowhere else). These factors, however, are often difficult to determine, given many plants' propagation through airborne seeds and their consequent tendency to travel, even without human assistance. Most of the plants that grew in the early modern European countryside were, in fact, actually a mixture of species which had spread across the continent since the end of the previous Ice Age; others which had gradually shifted their range due to climatic or environmental factors; still others which had originally reached an area through accident, perhaps a seed stuck to a traveler's clothing, but had now firmly established themselves in the new area; and yet others which had been introduced into Europe from elsewhere

---

[2] See for example A. G. Morton, *History of Botanical Science: An Account of the Development of Botany from Ancient Times to the Present Day* (New York: Academic Press, 1981), 197–8.

[3] Rudolph Zaunick, Kurt Wein und Max Militzer, eds., *Johannes Franke 'Hortus Lusatiae' Bautzen 1594, mit einer Biographie* (Bautzen: Naturwissenschaftliche Gesellschaft Isis, 1930), *passim* and Kurt Wein, "Die Wandlungen im Sinne des Wortes 'Flora,'" *Fedde, Repertorium Specierum novarum Regni vegetabilis, Beihefte* 66 (1932): 74–87. See also D. G. Frodin, *Guide to Standard Floras of the World*, 2nd ed. (Cambridge: Cambridge University Press, 2001), 26, for a balancing of the two claims. On the more recent career of local floras, see D. E. Allen, "Four Centuries of Local Flora-Writing: Some Milestones," *Watsonia* 24 (2003): 271–280 and Frodin, 27–46.

during the Middle Ages and Renaissance, some of them originally cultivated only in gardens, but now naturalized, i.e. to be found growing even in the wild.[4] Whether a species *is* or is not truly "indigenous" to a given area thus poses formidable conceptual problems today, even for the expert botanist.[5]

Nor is the phrase "local flora" itself of much help in clarifying this issue, since it did not come into use until the nineteenth century. Indeed, the word "flora" was in fact employed during the seventeenth and eighteenth centuries to refer to anything from the goddess Flora to the study of botany itself.[6] Likewise, though the plant catalogues that began to emerge in the early seventeenth century did indeed soon come to share many aspects of interior organization and convention, their exteriors belied their similarities, as authors creatively adorned their works' title-pages with practically every other term *except* "flora." A plant catalogue, for example, might be labeled a *hortus* (garden), a *viridarium* (plantation), or even a *museum*; alternatively, it might promise the reader anything from a mere *index* to full-fledged set of *deliciae* (delights). Given this situation, some commentators have declared Linnaeus in the mid-eighteenth century the true founder of the local flora, with his establishment of far more explicit and rigorous rules for the labeling and composition of local floras (using, of course, his own new system of nomenclature).[7]

In this chapter, I approach the origins of the local flora using the tools of the historian, rather than the botanist's contemporary reconstruction of individual species' modern identification. The focus here will thus be on *practices*, and in particular on how compilers of these works themselves, that is to say local florists, *articulated* their purposes and attempted to define their works' scope.[8] These changed over time. As we have seen, for example, while sixteenth-century authors of herbals and other botanical works drew increasingly on their own researches into local plants, they did not narrowly limit their geographical focus, nor present themselves as doing so. Though the study of local plant specimens indeed became a favorite activity of humanist physicians in Italy and then northern Europe, and they took copious notes,

[4] Hansjörg Küster, *Geschichte der Landschaft in Mitteleuropa: Von der Eiszeit bis zur Gegenwart* (Munich: Beck, 1995).

[5] See for example Mark W. Schwartz, "Defining Indigenous Species: An Introduction," in James O. Luken and John W. Thieret, eds., *Assessment and Management of Plant Invasions* (New York: Springer, 1997), 7–17.

[6] Janet Browne, *The Secular Ark: Studies in the History of Biogeography* (New Haven, CT: Yale University Press, 1983), 27–31; Wein, *passim*. For the sake of clarity, this chapter will use the term "flora" to refer primarily to the *genre* of the local plant catalogue in its printed form, rather than to the actual plants in an area.

[7] Browne, 27–31.

[8] Though the term "florists" is nowadays used mainly to refer to flower-sellers, it shall be used in this book to designate the compilers of local floras, early modern pursuers of Flora in the broader sense, that is to say the entire world of plants rather than just of flowers.

these notes were rarely published in their original form. Rather, the information was frequently reworked and reorganized, and inserted into texts
that more directly fulfilled encyclopedic humanist projects either of editing
and commenting on ancient authors, or of compiling works as comprehensive and compendious as possible.[9] As discussed in the previous chapter,
quite a few sixteenth-century authors, drawing on rhetoric similar to that of
Paracelsus, did indeed declare that their works would focus on "German"
plants, "French" ones, or those of other countries. But these grandiose claims
were not matched by any systematic efforts to cover such vast areas; for
example, it would have been almost unthinkable for any one individual to
attempt to locate and include every single plant to be found within the sprawl-
ing territories of the Holy Roman Empire, or any other large European state
for that matter. Only in the very last years of the sixteenth century did there
begin to appear plant catalogues explicitly declaring a more narrow, regional
focus; again, however, the contents of these works did not fully match their
ambitious titles (the German physician Johann Thal's 1588 *Sylva Hercynia*,
for example, did not come close to cataloguing all the plants within the Harz
forest its very title invoked).[10]

In the early seventeenth century, however, a series of plant catalogues
began to appear that, while likewise committing themselves to the study
of distinct geographical areas, now restricted their focus far more severely,
to that of a radius often of a mere several miles around a particular town.
These works all shared the distinctive format, described above, that would
become characteristic of the local flora. Strikingly small in size, yet rich in

---

[9] On humanist botanizing, see Karen Meier Reeds, "Renaissance Humanism and Botany,"
Annals of Science 33 (1976): 519–542; Karen Meier Reeds, *Botany in Medieval and Renaissance Universities* (New York: Garland, 1991); and Brian W. Ogilvie, *The Science of
Describing: Natural History in Renaissance Europe* (Chicago, IL: University of Chicago
Press, 2006). On occasion, though, works were published listing plants found on specific
journeys; these may be seen as important precursors to the local flora. In the English context, see for example Thomas Johnson's herborizing expedition in Kent, recorded in his *Iter
plantarum investigationis ergo susceptum a decem sociis, in agrum Cantianum* ([London?]:
A. Mathewes, 1629). Traces of these humanist botanizing practices remain in some local
floras; see for example the organization of Jacques Philippe Cornut, *Canadensium plantarum,
aliarumque nondum editarum historia. Cui adiectum est ad calcem enchiridion botanicum
Parisiense continens indicem plantarum, quae in pagis, silvis, pratis, & montosis iuxta Parisios* (Paris: apud Simonem Le Moyne, 1635).

[10] Johann Thal, *Sylva Hercynia, sive catalogus plantarum sponte nascentium in montibus
& locis plerisque Hercyniae Sylvae* (Frankfurt: apud Iohannem Feyerabend, 1588). For
similar examples from this period see for example Johann Wigand, *Vera historia de succino Borussico, de alce Borussica, & de herbis in Borussia nascentibus* (Jena: typis Tobiae
Steinmanni, 1590); Johann Francke, *Hortus Lusatiae* (Bautzen: excudebat Michael Wolrab.,
1594); Caspar Schwenckfeld, *Stirpium & Fossilium Silesiae catalogus* (Leipzig: impensis
Davidis Alberti, 1600); Pierre Richer de Belleval, *Dessein touchant la recerche des plantes du
pays de Languedoc* (Montpellier: Par Jean Gillet, 1605), reprinted in his *Opuscules* (Paris:
n.n., 1785).

information, they offered a densely compact web of knowledge about the places in which they had been produced. These places were distinctive in their own right; not only were they now mere towns, as opposed to entire territories, but, almost without exception, they were a very special kind of town: the college town. Local floras emerged in university towns large and small, from the famous trading center of Basel to tiny Altdorf itself. And they shared another feature in common: the universities in question were all among the earliest adopters of a crucial Italian innovation, the botanical garden, which had from its origins in Padua and Pisa begun to spread northwards over the Alps to sites like Montpellier and Leiden. By the early seventeenth century, formal botanical gardens attached to universities had begun to transform the European landscape from within. Now widely dispersed across the continent, they introduced not only new scents and colors wherever they were planted, but new forms of expertise and new kinds of interest in the natural world. It was within this specific set of circumstances that one highly trained social group, namely professors of medicine, came in the early seventeenth century to claim for themselves the task of defining and publicizing local nature, in particular the riches of plants and herbs to be found in local landscapes. They did this by creating the genre of the local flora.

The development of this particular form of local knowledge, in the form it took, was thus a highly contingent one. It could have taken other paths. For many different forms of natural knowledge relating to local environments were available to early modern Europeans. To cite only a few examples, peasant agricultural knowledge, the pharmaceutical skills of apothecaries, and the plant-collecting of midwives and "wise women" offered numerous different ways of framing and using knowledge about local plants.[11] Other institutions, such as courts and scientific societies, also offered attractive venues for the advancement of natural knowledge.[12] And outside Europe,

---

[11] On the latter, see for example Katharine Park, "Medicine and Magic: The Healing Arts," in *Gender and Society in Renaissance Italy*, ed. Judith Brown and Robert Davis (London: Longman, 1998), 129–149; Margaret Pelling and Charles Webster, "Medical Practitioners," in *Health, Medicine and Mortality in the Sixteenth Century*, ed. Charles Webster (Cambridge: Cambridge University Press, 1979), 182–184, 233–235; and, on the controversial question of the herbal knowledge of accused witches, Sergius Golowin, *Die Weisen Frauen: Die Hexen und ihr Heilwissen* (Basel: Sphinx, 1982) and Gunnar Heinsohn and Otto Steiger, *Die Vernichtung der Weisen Frauen* (Herbstein: März, 1985). On agricultural knowledge of plants, see Mauro Ambrosoli, *The Wild and the Sown: Agriculture and Botany in Western Europe, 1350–1850* (Cambridge: Cambridge University Press, 1997).

[12] See for example Bruce T. Moran, "Patronage and Institutions: Courts, Universities, and Academies in Germany; an Overview 1550–1750," in *Patronage and Institutions: Science, Technology, and Medicine at the European Court, 1500–1750*, ed. Bruce T. Moran (Woodbridge, UK: Boydell, 1991), 169–184; and Mario Biagioli, *Galileo, Courtier: The Practice of Science in the Culture of Absolutism* (Chicago, IL: University of Chicago Press, 1993).

of course, still further forms of natural knowledge flourished.[13] Yet what happened in the case of the local flora was that as the genre became codified and standardized within Europe, it simultaneously became the province of a small and select group of people: academic physicians. By choosing to write and publish local floras, and to do so in the ways they did, they effectively came to determine what would count as genuine and authoritative information about the local plant worlds they described. In this way, their particular form of what might be called "local knowledge" ended up becoming accepted as *the* valid botanical "local knowledge" for and of places everywhere. By the mid-eighteenth century, when Linnaeus appeared on the scene, a pattern had been set that would endure: while local floras following standardized formats were only just beginning to be written for the colonies, their European counterparts had already succeeded in taking inventory of the vast majority of Central and Western European plant species. These parts of Europe had now acquired the status of being, on the whole, quite well "known" to and by botanical science; the rest of the world would be seen as lagging behind.

FIELD TRIPS

Making one's way along the narrow streets of Altdorf today, one would scarcely imagine that this small town, barely eight blocks in diameter inside its old walls, had once been home to a lauded university, or that its countryside's plant life had for several centuries been among the most thoroughly investigated of any in Europe. For present-day visitors disembarking at the tiny commuter-rail platform, the rural nature of the town presents itself immediately; open fields abut one side of the tracks, while on the other are posted maps of nearby hiking trails for twentieth-century environmentalists and nature-lovers. A tidy suburb street leads obliquely to one of the town's surviving gates, and to the main street, where a stationery shop sells postcards of the town church to the occasional tourists who pass through. An exhibit in an archway reveals that the former university, closed in the Napoleonic Wars, has been turned into a group home for people with disabilities. The former botanical garden, built just outside the town walls, is no more – it has vanished into a modest suburb where, however, chickens still roam the yards. Yet this unassuming town ended up producing more accounts of its botanical surroundings than almost any other in Europe. Why? This section will focus on the "local" example of Altdorf in the century-and-a-half before its garden party in 1727, as a lens through which to view the origins

---

[13] To attempt to cite the full range of scholarship on this topic would be overwhelming; for an introduction, see Helaine Selin, *Nature across Cultures: Views of Nature and the Environment in Non-Western Cultures* (Dordrecht: Kluwer, 2003).

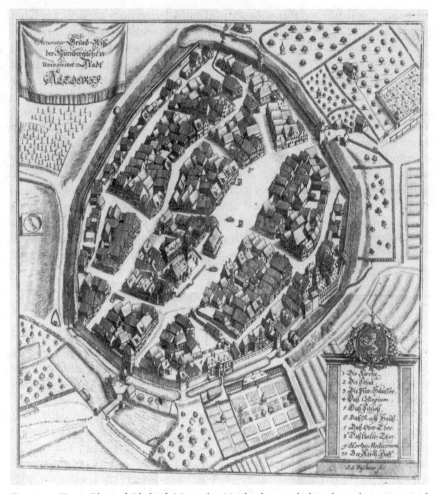

Figure 4. Town Plan of Altdorf. Note the *Marktplatz* and church at the center, and the botanical garden just outside the city walls at the lower center-right. From Johann Jakob Baier, *Ausführliche Nachricht von der Nürnbergischen Universität-Stadt Altdorff* (Nuremberg, 1717). Courtesy of the Harvard College Library.

and development of the local flora as a genre in the seventeenth century. For the case of Altdorf in fact illustrates many common themes in the rise of the local flora in early modern Europe. By tracing the history of the local flora in Altdorf, and pointing out the similarities and differences between its case and others, this section will explore the emergence of this small, yet powerful, tool for taking inventory of the natural world.

There are a number of points at which one could begin this story – the foundation of the town of Altdorf itself within some indeterminate medieval period, for example, or the more precise date of 1504, when, after bravely defending itself for four days and nights, the small town surrendered to the armies of the free imperial city of Nuremberg and became part of that city's expanding territory.[14] But as good a place as any to start is with the decision in 1575 of a group of Nuremberg burghers to found an academy for the education of their sons, and to situate it outside the city in the country town of Altdorf. The sixteenth century had seen the foundation of numerous such academies, as the Protestant–Catholic divide and subsequent confessionalization had spurred the multiple foundings of new institutions.[15] But Nuremberg's decision to locate its new academy in the countryside was not a particularly obvious decision to make. Nuremberg, as a well-established center of trade with a rich assortment of community and religious institutions, could easily have afforded and accommodated an academy within its own walls. But other concerns were on the minds of committee members. After considering a number of sites, including buildings within the city walls, they rejected them all in favor of Altdorf. Altdorf's rural situation was what had swayed the committee. Out in the countryside, they hoped, the new academy might provide a peaceful and quiet atmosphere for their sons, far enough from the temptations of city life (see Figures 4 and 5).[16]

The potential the Altdorf countryside offered for botanizing was, however, certainly not at the top of anyone's list of considerations. Although the newly-founded academy was relatively well-funded, owing to the generosity of the initial donors, it was still a provincial academy. It employed two medical professors, in itself a remarkable achievement for an academy that did not yet possess the ability to grant medical degrees, but neither of these professors had a special interest in botany. Since the academy could not grant doctorates, only bachelor's degrees, instruction in many topics was limited.[17] Though many students passed through town, few remained for long. After learning the basics of the liberal arts, a student could indeed proceed to study theology, philosophy, law, and medicine, but if he wanted a *degree* in any of these areas he had to travel to other nearby universities, in the peripatetic style of the early modern student. Or he might travel to Italy, as students from all over the European continent increasingly

---

[14] Heinz Dannenbauer, *Die Entstehung des Territoriums der Reichstadt Nürnberg* (Stuttgart: Kohlhammer, 1928), 4–7 and 182–183; Wolfgang Leiser, "Das Landgebiet der Reichsstadt Nürnberg," in *Nürnberg und Bern: Zwei Reichsstädte und ihre Landgebiete*, ed. Rudolf Endres (Erlangen: Universitätsbibliothek Erlangen, 1990), 227–260.

[15] Hilde de Ridder-Symoens, ed., *A History of the University in Europe, vol. 2: Universities in Early Modern Europe, 1500–1800* (Cambridge: Cambridge University Press, 1996), 71.

[16] Emil Reicke, *Geschichte der Reichsstadt Nürnberg* (Nuremberg: Raw, 1896), 940; Horst Recktenwald, "Aufstieg und Niedergang der Universität Altdorf," in *Gelehrte der Universität Altdorf* (Nuremberg: Lorenz Spindler Verlag, 1966), 14–15.

[17] Recktenwald, 17; see also Steven Ozment, *Three Behaim Boys: Growing up in Early Modern Germany* (New Haven, CT: Yale University Press, 1990).

Figure 5. The *Collegium* (main university building) at Altdorf. This building housed the university's lecture halls. From Johann Jakob Baier, *Ausführliche Nachricht von der Nürnbergischen Universität-Stadt Altdorff* (Nuremberg, 1717). Courtesy of the Harvard College Library.

did.[18] In his academic peregrinations, the famous philosopher Leibniz himself passed through Altdorf, picking up a law degree along the way.[19]

Yet during the earliest years of the seventeenth century, two medical students turned the Altdorf countryside – a sleepy place of meadows and rolling hills – into a site for research in and of itself. The two were Ludwig Jungermann and Caspar Hofmann. Jungermann was the son of a professor himself, a circumstance not at all rare in the nepotistic and family-structured world of the early modern university. His father taught law in Leipzig, where he had also held the ceremonial positions of Senior and Vice-Chancellor, and (seven times!) Rector of the university.[20] Hofmann, in contrast, was the son of a smith, and had only reached the university level thanks to the intercession of a school administrator in Gotha and the awarding of a series of scholarships.[21] Both had reached Altdorf in the 1590s after a succession of academic

[18] See for example Loris Premuda, "Die Natio Germanica an der Universität Padua," *Sudhoffs Archiv* 47, 2 (1963): 97–105.
[19] E. J. Aiton, *Leibniz: A Biography* (Bristol, UK: Adam Hilger, 1985), 22.
[20] Georg Andreas Will, *Nürnbergisches Gelehrten-Lexicon* (Nuremberg: zu finden bei L. Schupfel, 1755–1758), II, 261–262.
[21] Will, II, 162.

peregrinations, which had taken them to Jena and Strasbourg respectively. In Altdorf they studied under an assortment of professors, including Nicolaus Taurellius and Philippus Scherbius, who together comprised the academy's entire medical faculty at the time.[22] Although the two youths were in fact only several months apart in age, Hofmann was the more senior student, having already spent several years in Padua extending his knowledge of medicine at one of Italy's most respected universities.[23] Jungermann soon met his more experienced counterpart, with whom, as an eighteenth-century Nuremberg antiquary tells us, "he formed a close acquaintance, and began to collect the local plants and herbs."[24]

Here is where the evidence ends. Neither the notes themselves, nor any correspondence from this period of Jungermann's and Hofmann's relationship survive to let us know when, and why, the two began their floristic endeavors. For the study of the local plant world was not yet fully institutionalized in Altdorf – nor, for that matter, in most other comparably-sized academies, even in Italy where the academic study of botany was most advanced. Though universities across Europe had begun to require that professors "demonstrate" to students "simples," or medicinal herbs, as well as the anatomy of human bodies, such teaching could still be rare and/or only informally conducted.[25] It partly depended on local institutional arrangements. Just as the anatomy theater was to become the quintessential early modern site for the performance of dissection, so too would the botanical garden, imported from Italy, become essential for the demonstration of plants. With its concentration of a wide variety of species into a minimal and ordered space, the garden served as a "living classroom" for students; in a broader sense, it also can be seen as forming a creative, even utopian space in which natural objects could be manipulated and thus take on new symbolic arrangements.[26] If a university did not yet have its own "public" garden, the private gardens of individual professors, or of local elites, might be used. It is likely that Hofmann and Jungermann availed themselves of some of the latter; by the late sixteenth century, Nuremberg had already become known for the private

[22] Dorothee Flessa, "Die Professoren der Medizin zu Altdorf von 1580–1809," Diss. Med., Universität Erlangen, 1969, 7–12; Magnus Schmid, "Medici Universitatis Altorfinae," in *Gelehrte der Universität Altdorf*, ed. Horst Claus Recktenwald (Nuremberg: Lorenz Spindler Verlag, 1966), 79.

[23] Elias von Steinmeyer, *Die Matrikel der Universität Altdorf* (Würzburg: Königliche Universitätsdruckerei H. Stürtz, 1912), I, 54 and 62.

[24] Will, II, 262.

[25] Reeds, *Botany in Medieval and Renaissance Universities*, 41–64 and 110–111.

[26] John Prest, *The Garden of Eden: The Botanic Garden and the Re-Creation of Paradise* (New Haven, CT: Yale University Press, 1981) and Richard Patterson, "The 'Hortus Palatinus' at Heidelberg and the Reformation of the World," *Journal of Garden History* 1 (1981): 67–104 and 179–202.

gardens of its learned citizens.[27] But in its Academy, botanical activities, and in particular field trips, did not yet constitute part of the standard medical curriculum. Jungermann and Hofmann, in traversing the local landscape, did so without significant institutional support.

Eventually, Hofmann's and Jungermann's botanical friendship was to be cemented in the form of a book, a "catalogue of plants around Altdorf near Nuremberg, and in some neighboring places" (1615).[28] The publishing history of this work is complicated, but rewarding in its details. Basically, the paths of the two men parted during the 1610s, as Jungermann moved away from Altdorf to get his doctorate in Giessen, where he ultimately assumed a professorship. Hofmann, meanwhile, stayed in Altdorf and took over Taurellius's chair in the medical faculty when the latter died. But a noteworthy event brought the two former classmates back together again in 1613: Jungermann's travel to the area to assist in the production of the *Hortus Eystettensis*, a massive and sumptuously illustrated volume of garden plants commissioned by the wealthy prince-bishop of Eichstädt, not far to the south.[29] At some point during Jungermann's travels to and from Eichstädt he and Hofmann met up again, and agreed to take the significant step of actually *publishing* their notes on the Altdorf flora. Since Jungermann was otherwise occupied, both at Eichstädt and back in Giessen, Hofmann ended up taking on the lion's share of the editing. For the book's publication in 1615, Hofmann wrote a preface clarifying his and his partner's respective roles; but it was Jungermann's name that appeared first on the title page.[30]

Several points are worth mentioning here. Ultimately, the catalogue of the Altdorf flora needed two people's work to bring to completion. The project of the local flora thus ended up being, in this case and subsequently in many others, a collaborative one. This tension between individual authorship and group responsibility was to persist in later local floras, owing to

---

[27] The Nuremberg scholar Joachim Camerarius, for example, was renowned in the late sixteenth century for his private garden in Nuremberg; see his *Hortus medicus et philosophicus* (Frankfurt: apud Hannem Feyerabend, 1588) and, on his fruitful relationship with Italian naturalists, Giuseppe Olmi, "'Molti amici in varii luoghi': Studio della natura e rapporti epistolari nel secolo xvi," *Nuncius* 6 (1991): 3–31.

[28] Ludwig Jungermann, *Catalogus plantarum, quae circa Altorfium Noricum, et vicinis quibusdam locis nascuntur* (Altdorf: apud Cunradum Agricolam, 1615).

[29] Basil Besler, *Hortus Eystettensis* ([Nuremberg]: n.n., 1613). See Hans-Otto Keunecke, ed., *Hortus Eystettensis: Zur Geschichte eines Gartens und eines Buches* (Munich: Schirmer-Mosel, 1989).

[30] It was himself, in fact, not Jungermann, to whom Hofmann gave the Latin title of *"autor"* of the book (Jungermann, sig. A2v); by doing so, though, Hofmann would not have meant to imply that the work was primarily his own creation, in the modern sense, but rather that he had played a major role in bringing the work to physical publication. On changing early modern understandings of authorship in the academic context, see Gertrud Schubart-Fikentscher, *Untersuchungen zur Autorschaft von Dissertationen im Zeitalter der Aufklärung* (Berlin: Akademie-Verlag, 1970).

the enormous amounts of "labor" required by the work of empirical natural history. Ultimately, the local flora would be the product of countless botanical excursions, of an (all-male) group activity carried out over an extended period of time.[31] Likewise, as local floras came to be written repeatedly about particular areas, compilers usually drew at least in part on the efforts of their predecessors. Though conservative university medical faculties operated in an environment rather different from Bacon's utopian dreams of "many hands making light work," the amount of work involved in compiling local floras would, in fact, be quite considerable. This work might be largely hidden on the title page, where the responsible medical professor would be given credit; but it would be there nonetheless. The local flora thus became a group enterprise as much as an individual one.

Here another issue must be raised: that of local knowledge. For neither Jungermann nor Hofmann were, in fact, *locals* of Altdorf, or even of nearby Nuremberg. The towns they both had grown up in were more than a hundred miles away, at least several days' journey by horse, more by foot. In the intensely parochial world of early modern Europe, this meant that Jungermann and Hofmann, on their first arrival in Altdorf, would have been considered "foreigners." Upon being awarded their university posts, they were indeed then granted a special status within the town, but in many ways they would still have been regarded as outsiders in it. With the two learned physicians' history of mobility, they can be seen as exemplifying what one scholar has labeled the "movers and doers" of German society, set apart from and often resented by far-less-mobile "hometownsmen" and "countrymen."[32] Like Jungermann and Hofmann, very few authors of local floras had actually grown up in the areas they wrote about. Rather, they were often recent arrivals, who might well move on once again in several years if summoned to a post at another university, as was the case with Jungermann himself. The tendency of professors to try to pass on their positions to their sons or sons-in-law, forming what were effectively academic dynasties, did mean that in some cases compilers of local floras had a more longstanding and intimate acquaintance with the areas they undertook to describe; yet even in these situations, the broader experience of the world they eventually gained through academic peregrinations abroad set them apart from other

[31] On women's general lack of admission to university medical study (or any other form of university study, for that matter) during the early modern period, see Londa Schiebinger, *The Mind Has No Sex? Women in the Origins of Modern Science* (Cambridge, MA: Harvard University Press, 1989). There were some exceptions, for example in Italy, but even there it was very rare for a woman to be permitted entrance: see Paula Findlen, "The Scientist's Body: The Nature of a Woman Philosopher in Enlightenment Italy," in *The Faces of Nature in Enlightenment Europe*, ed. Lorraine Daston and Gianna Pomata (Berlin: BWV, 2003), 211–236.
[32] For a discussion of these categories, see Mack Walker, *German Home Towns: Community, State, and General Estate 1648–1871* (Ithaca, NY: Cornell University Press, 1971), 119–133.

townspeople.[33] Even while university professors took on the role, then, of "representing" the natural world of the towns they now lived in, by publishing local floras of and for them, the perspectives they brought to the task tended to be quite different than those of most long-term inhabitants. While authors of local floras drew on the experience of such inhabitants as informers whenever possible, the documents they produced were far from mere transcriptions of such genuinely local knowledge, whose providers were, in any case, only very rarely cited. The "local knowledge" set down by authors of local floras, while bearing clear evidence of their devotion to their adopted towns, was not primarily intended for the use or benefit of the townspeople themselves. It had another audience.

And this leads to one final point. Caspar Hofmann, in the preface to his and Jungermann's local flora of Altdorf, made the primary audience of his book extremely clear: medical students, some of the most peregrinatory individuals in all of early modern Europe. Struck by their ignorance of medicinal herbs, Hofmann wrote, he had decided to "lead them out often into the broadest garden of Nature, into our fields and even our forests, so that they might themselves inspect and learn to recognize those plants, which they would one day use."[34] In short, Hofmann came to establish the plants of the Altdorf countryside – that "broadest garden of Nature" – as his province for the instruction of the "movers and doers," a site for his official duties as a representative of the Altdorf Academy. Whereas earlier botanizing at Altdorf had been more of an independent or leisure activity, to be carried out on one's own, Hofmann claimed the Altdorf countryside as an extension of the academy's own facilities. And, as a matter of fact, through the species he listed, Hofmann claimed not just the countryside's medicinal plants, but *all* of its vegetation; as would be the case with almost all local floras, Hofmann's catalogue in fact included a wide variety of plants *not* especially known for medicinal use, but simply common in the area. By thus making the local nature of Altdorf public through the publication of its inventory, Hofmann effectively annexed its fields and forests as sites for the pursuit not only of genuinely local interests, but of natural knowledge more generally. The knowledge he set down would, in fact, be in a format scarcely accessible to locals themselves; only those who were already effectively outsiders, in one form or another, would be able to make any sense of it.

---

[33] On academic dynasties, see Friedrich W. Euler, "Entstehung und Entwicklung deutscher Gelehrtengeschlechter," in *Universität und Gelehrtenstand 1400–1800*, ed. Helmuth Rössler and Günther Franz (Limburg: Starke, 1970), 183–232; William Clark, *Academic Charisma and the Origins of the Research University* (Chicago, IL: University of Chicago Press, 2006), 242–243; and Alix Cooper, "Homes and Households," in *The Cambridge History of Science, Vol. 3: The Early Modern Period*, ed. Katharine Park and Lorraine Daston (Cambridge: Cambridge University Press, 2006), 224–237.

[34] Jungermann, sig. A2r.

The story of Altdorf's flora, though, was not yet over. In 1622 the situation changed: with the approval of the Emperor in Vienna, and as a result of complicated maneuverings in the Imperial court, the small academy was granted university status.[35] The new university took several immediate steps. First, it summoned Ludwig Jungermann back to Altdorf, where in 1625 he assumed a newly-created third medical professorship. A specific task came with the post: he was to create and open a botanical garden, as the new university's first official scientific institution.[36] As an institution operated by the university itself, the new botanical garden would be "public," in contradistinction to the private gardens students had previously relied upon. Furthermore, the botanical garden would come to symbolize the universality of the new institution, with its "exotic" plants brought in from all four corners of the globe.[37] Jungermann fulfilled this task with astonishing speed, opening the new *"hortus publicus"* to the public in 1626, a scant four years after the academy's elevation. From this point on, the botanical garden, with its abundance of exotic plants, would be the focus for Altdorf's botanical activities. Far from eclipsing local nature, though, it was to serve as a center from which local nature might be even more closely scrutinized.

It took Jungermann nine more years to accumulate enough exotic plants to publish his first catalogue of the garden.[38] The work of opening the botanical garden, of laying out the ground and hiring gardeners, had, in contrast, been easy; what took time was the slow accumulation of enough exotic specimens to rival other botanical gardens in the Empire and abroad. This was a lengthy process. Jungermann used his connections in Nuremberg, and Nuremberg's strategic location as a trading center, to procure plants and seeds from a wide range of foreign sources. In particular, he acquired plants from the Italian cities with which Nuremberg had long traded, with their Mediterranean contacts. He also turned to those Dutch gardeners and *"Materialisten"* with whom Nuremberg merchants had been cultivating contact. With their growing colonial empire, the Dutch could supply seeds from the Americas, from Africa, and from points all over the Far East. Jungermann planted the seeds and waited. Eventually, in 1635, he published the catalogue of those plants that had survived the German winters.

Jungermann's next effort placed exotic and local plants side by side, in his catalogue of the plants to be found both "in the medical garden and the

[35] Anton Ernstberger, "Die feierliche Eröffnung der Universität Altdorf (29. Juni 1623)," in *Franken – Böhmen – Europa* (Kallmünz, 1959), 156.
[36] Heinz Röhrich, "Zur Geschichte des 'Doctorgartens' oder 'Hortus medicus' der ehemaligen Nürnberger Universität Altdorf," *Erlanger Bausteine zur fränkischen Heimatforschung* 11 (1964): 31–42.
[37] Frans A. Stafleu, "Botanical Gardens Before 1818," *Boissiera* 14 (1969): 31–46.
[38] Ludwig Jungermann, *Catalogus plantarum, quae in horto medico Altdorfino reperiuntur* (Altdorf: typis Balthasari Scherffi, 1635).

Altdorf countryside."³⁹ In the preface to this work, he declared his intention to devote his volume equally to "matters both of the garden and of
the forest" (*tam hortensium ... quam sylvestrium*).⁴⁰ This complementarity
reappears throughout the volume, for example amongst several pieces of
poetry attached to the front. One four-line verse was titled "to the medical garden and the plants growing of their own accord (*spontanea*) in Altdorf."⁴¹ Another inquired: "You seek compendia of both field and garden?
The page assigns both to medical purposes."⁴² In this work, then, we can
see a sort of parallelism. For Jungermann, exotic plants in gardens and their
local counterparts in surrounding fields were clearly at opposite poles of
a rhetorical opposition, yet they were simultaneously linked by this very
opposition.

But it is from the works of Jungermann's successor that we first get a true
glimpse into the experience of local flora-writing at Altdorf, and the various practices – such as that of the field trip – that lay behind it. For Moritz
Hoffmann (despite his name, no relation to Caspar Hofmann) published not
only garden catalogues and local floras, but also several revealing guides to
the techniques of the local botanical excursion. For Hoffmann, as for Jungermann, the local and the exotic were clearly parallel. Hoffmann published his
Florae Altdorfinae in two parts, one of "garden delights" (*Florae Altdorfinae
deliciae hortenses*) and the other of "forest delights" (*Florae Altdorfinae deliciae sylvestres*), containing respectively a catalogue of the botanical garden
and a catalogue of local plants.⁴³

Each of these volumes of "flowers of Altdorf" represented the town
and university, no less than the other. The garden catalogue, for example,
was dedicated to eight men, all of whom were councilors (*Duumviris* and
Septemviris) and/or administrators (*Scholarchis*) of both Nuremberg and the
university. Representatives of many of the patrician families of Nuremberg
were on his list of dedicatees (the Behaims, Harßdörffers and Welsers, for
example), displaying the connections between city and university. Hoffmann
thanked them for their prudent efforts "on behalf of justice, on behalf of
the general health of the citizens, on behalf of the entire Republic and
Academy."⁴⁴ Just as Atlas supported the heavens, so too, claimed Hoffmann,
did these patrons support the standing of their city ("you hold up the skies

³⁹ Ludwig Jungermann, *Catalogus plantarum, quae in horto medico et agro Altdorphino reperi-
    untur* (Altdorf: typis R. Scherffi, 1646).
⁴⁰ Jungermann, *Catalogus plantarum ... in horto medico et agro*, sig. [A1]v.
⁴¹ Jungermann, *Catalogus plantarum ... in horto medico et agro*, sig. [A4]v.
⁴² Jungermann, *Catalogus plantarum ... in horto medico et agro*, sig. [A4]v.
⁴³ Moritz Hoffmann, *Florae Altdorfinae deliciae hortenses sive Catalogus plantarum horti
    medici* (Altdorf: typis Georgi Hagen, 1660); Moritz Hoffmann, *Florae Altdorfinae deliciae
    sylvestres sive catalogus plantorum in agro Altdorffino locisque vicinis sponte nascentium*
    (Altdorf: typis Georgi Hagen, 1662).
⁴⁴ Hoffmann, *Florae Altdorfinae deliciae hortenses*, sig. [i]v.

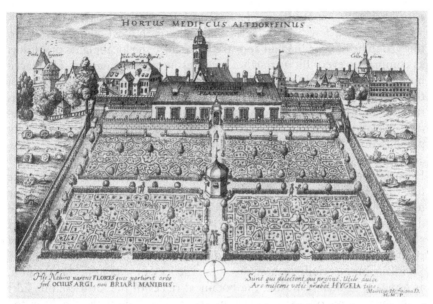

Figure 6. The Botanical Garden at Altdorf. Visitors stroll through the garden, and one of them points at the plants. Note at the top center the greenhouse where tropical plants were nurtured, and on both left and right the neat rows of fossils (obviously not to scale). From Moritz Hoffmann, *Florae Altdorfinae deliciae hortenses* (Altdorf, [1660]). Library of the Arnold Arboretum, Harvard University, Cambridge, Massachusetts, USA.

of your fatherland").[45] In particular, he cited their recent funding for Altdorf of both an anatomy theater (the *Theatrum Anatomicum*) and a greenhouse for exotic plants, which he labeled in turn a "Botanical Theater" (*Theatrum Botanicum*). These gifts, he suggested, put them in the good company of the "princely men and magnates throughout Europe" who had likewise set aside space for "Anatomical and Botanical Theaters" for the greater glory of their communities.[46] Altdorf's botanical garden, with its catalogue, thus served as a theater for showing off the best that the University (and its patrons) could do.

Yet just as the "garden delights" of Altdorf were crucial for university prestige, so too were its "forest delights," maintained Hoffmann. This may be seen visually, for example, in an engraving of the Altdorf *hortus medicus* Hoffmann had apparently commissioned for the garden catalogue, depicting a series of local fossils which he had presumably found on his botanical excursions, lying neatly numbered and in rows directly outside the walls of the Altdorf garden (see Figure 6). He dedicated his *Florae Altdorffinae*

---

[45] Hoffmann, *Florae Altdorfinae deliciae hortenses*, sig. [i]v.
[46] Hoffmann, *Florae Altdorfinae deliciae hortenses*, sig. [i]r.

deliciae sylvestres to an almost identical group of notables as his earlier effort, adding merely a few "Provincials," or nobles with local country seats. To this group he now penned an appeal to recognize the local plants of Altdorf as comprising every bit as much a local resource as the garden. He placed his discussion of the local flora's virtues within the crucial framing story of the indigenous versus the exotic. In the brief history of medicine he gave his readers, Europeans had at some point turned away from "domestic remedies, easy to prepare," in favor of "overseas ones, ambitiously and carefully purchased from the Indies."[47] This latter "farrago of medicaments," "new and unused, sought from another world," had replaced familiar remedies.[48] From his history, Hoffmann drew the conclusion that it was time for the tide to turn the other way: that far from spurning local plants, it was time to recognize how much more suitable they were, not only for curing local diseases, but for preventing them in the first place. In the region of Altdorf alone, Hoffmann was proud to report, he had found over 900 species of plants "no less elegant than commendable for Medical virtue."[49]

Hoffmann especially commended the practice of the botanical excursion, as a means of demonstrating the botanical riches of the local countryside. Only these forms of "experience," he stressed, namely those taught in the university, could ensure an expert knowledge. Here he injected a threatening note, warning that without this knowledge, the populace might eat poisonous plants unawares.[50] It was for this reason that Hoffmann had originally instituted botanical demonstrations in the garden; and that he had eventually expanded these to include "not just Herbations but frequent excursions [with students] to higher and herbaceous places" within a boundary of three miles of Altdorf.[51] Hoffmann thus presented the botanical excursion as an explicit route towards the gaining of a new kind of knowledge, one which would exclude the "vulgar" or uneducated. As an added incentive, Hoffman argued that during his excursions, he had been able to find entirely new species, previously unknown. The local botanical excursion thus allowed the university physician to claim the surrounding countryside as his own.

---

[47] Hoffmann, *Florae Altdorfinae deliciae sylvestres*, sig. [i]v.
[48] Hoffmann, *Florae Altdorfinae deliciae sylvestres*, sig. [i]v.
[49] Hoffmann, *Florae Altdorfinae deliciae sylvestres*, sig. [i]r. When contemplating the figure of 900 species claimed here, one needs to bear in mind that not all plants that were listed may actually have been "medicinal" plants per se, and that species definitions at the time varied in their scope, i.e. what might be seen nowadays as merely different *varieties* of the same species were often viewed in the early modern period as entirely separate species.
[50] Hoffmann, *Florae Altdorfinae deliciae sylvestres*, sig. [i]r.
[51] Hoffmann, *Florae Altdorfinae deliciae sylvestres*, sig. [i]r. For a history of botanical excursions, though primarily within the English context, see D. E. Allen, "Walking the Swards: Medical Education and the Rise and Fall of the Botanical Field Class," *Archives of Natural History* 20 (1993): 335–367.

Hoffmann left behind an extremely revealing document about the practice of fieldwork in the botanical excursion: his *Botanotheca*. Originally conceived as a guide to the making of herbaria, this brief treatise, published in 1662, introduces the reader to the wide range of techniques necessary for successfully exploring a local landscape and returning with its bounty.[52] "Let's consider Botany as an operative art," Hoffmann proposed, and went on to list the tools and instruments needed to carry out a successful expedition in the "fields and somewhat ignoble places" of the countryside.[53] In addition to his "*Rhizotomus*," literally "root-cutter," as his primary tool, the aspiring botanist would also need a shoulder bag in which to carry his "*Torcularium Botanicum*" (a plant-press for preserving specimens), theriac (medicine against snakebite), spices (to purify the water en route), plus a sundial ("*Sciathericum, seu horologium solare*," so that he would always know the time). Last but not least, he should bring along a book about plants; Hoffmann suggested several of the classic herbals in pocket editions ("*minori forma oblonga*") as well as a "catalogue of plants growing spontaneously around Altdorf" (presumably either his own, which had just been published, or Jungermann's earlier effort).[54] Although it is unclear here whether Hoffmann thought the Altdorf guide might provide a useful checklist for other areas as well, in any case the botanical book, and in particular the local flora, emerges as a "tool" in its own right. Hoffmann proceeded to give further advice to the excursionist, for example that it was best to go plant-hunting at noon, since by then the early morning dew would have evaporated and any plants found would be drier (*sicciores*) and more easily inserted in the *hortus siccus*.[55] Subsequent topics included how to dry the plants, how to order them, and how to glue them to the page: "the most important thing about glue is, that it should possess the power to stick together."[56]

One final, and even more rewarding, glimpse into the world of the botanical excursion is offered by a book which Hoffmann published towards the end of his career, presenting the "medico-botanical description" of a mountain near Altdorf, the Moritzberg. Subtitling this work a "catalogue of plants offering themselves in botanical excursions" (i.e. to the mountain), Hoffmann used this book to recount the story of one particular excursion he took with his students up, down, and around the mountain.[57] He dedicated

---

[52] Moritz Hoffmann, *Botanotheca Laurembergiana* (Altdorf: typis Georgi Hagen, 1662).
[53] Hoffmann, *Botanotheca*, sig. [A2]v and A3r.   [54] Hoffmann, *Botanotheca*, sig. [A3]r.
[55] Hoffmann, *Botanotheca*, sig. [A4]r.
[56] Hoffmann, *Botanotheca*, sig. [D4]r. On the development of techniques for assembling herbaria, see Brian W. Ogilvie, *The Science of Describing: Natural History in Renaissance Europe* (Chicago, IL: University of Chicago Press, 2006), 165–174.
[57] Moritz Hoffmann, *Montis Mauriciani in agro Leimburgensium, medio inter Norimbergam & Hirsbruccum* (Altdorf: typis Henrici Meyeri, 1694).

the book to a Nuremberg city councilor, Karl Gottlieb Fürer, whose family seat was located on the mountain and through whose kind permission he and his students had been allowed to roam freely.[58] He loaded his book with praises both of the family and of the mountain, which he compared to that of Monte Baldo near Verona, renowned among Italian naturalists for its rare flora and possibilities for excursions.[59] After brief histories of Altdorf, Nuremberg, and the university itself, he launched his story with an image of medical students sallying forth upon their journey: "To the medical students leaving by the upper gate through the town ditches, the following plants present themselves. . . . "[60] Here he gave a long list of plants, the first of many that were to follow over the course of the book, as he detailed the plants of place after place. The remainder of the book followed Hoffmann and his students, as they slowly made their way to the mountaintop, where they found a memorial set up by the generosity of the Fürer family; and where they could view all four corners of the countryside, through which they now descended.[61]

In Hoffmann's guide to the Moritzberg, we see him using the botanical excursion as a way to extend the knowledge of local plants beyond the immediate area of Altdorf. Hoffmann declared the Moritzberg to be "just like a public garden, where common and rare plants, useful to those citizens protecting health, ought to be inspected every year."[62] By moving his botanical efforts out beyond the immediate surroundings of the university, into the landscape beyond the town, Hoffmann not only increased the size of his catalogue of plants, but also went a good way towards making this landscape "public."[63] Just as his predecessors Ludwig Jungermann and Caspar Hofmann had sought to present local lands as a resource for the university, arguing that they were every bit as useful a source of plants as the local botanical garden, so too did Moritz Hoffmann annex the local countryside

---

[58] Hoffmann, *Montis Mauriciani*, sig. [I]r.
[59] Hoffmann, *Montis Mauriciani*, 3. On early botanizing around Monte Baldo, see Francesco Calzolari, *Il viaggio di Monte Baldo, della magnifica città di Verona* (Venice: Vincenzo Valgrisio, 1566), and Paula Findlen, *Possessing Nature: Museums, Collecting, and Scientific Culture in Early Modern Italy* (Berkeley: University of California Press, 1994), 180–182.
[60] Hoffmann, *Montis Mauriciani*, 4.   [61] Hoffmann, *Montis Mauriciani*, 9–12.
[62] Hoffmann, *Montis Mauriciani*, sig. [i]r.
[63] On the significance of natural inquiry becoming seen as a public matter, rather than one to be conducted within the relative secrecy of craft and guild traditions, see William Eamon, *Science and the Secrets of Nature: Books of Secrets in Medieval and Early Modern Culture* (Princeton, NJ: Princeton University Press, 1994) and Pamela O. Long, *Openness, Secrecy, Authorship: Technical Arts and the Culture of Knowledge from Antiquity to the Renaissance* (Baltimore, MD: Johns Hopkins University Press, 2001). On the idea of a "public culture" of science, though in a somewhat later period, see Jan Golinski, *Science as Public Culture: Chemistry and Enlightenment in Britain, 1760–1820* (Cambridge: Cambridge University Press, 1992).

for the use of university scholars.[64] Arguing that the inspection of plants on the Moritzberg ought to be turned into an annual university ritual, Hoffmann in effect sought to turn the mountain into a form of community institution, just as the local countryside around Altdorf, thanks to his and his predecessors' efforts, had already become.

By 1727, the year in which the Altdorf botanical garden celebrated its centenary, the countryside around Altdorf had been canvassed and explored in half a dozen catalogues and numerous expeditions. As generation after generation of scholars had set forth from the Altdorf gates, a tradition of floristics had been established, which had uncovered plant after plant in place after place around the town, till few remained to be found. This, indeed, was Baier's complaint in 1727, as he surveyed the publication efforts of the Altdorf medical faculty for his centenary volume. It might be time, he conceded, for a new edition of garden catalogue and local flora, presenting them both in a single volume; he lamented that he had not had the time to do this himself.[65] As for writing a completely new local flora, that was out of the question; Moritz Hoffmann, lamented Baier, had done far too good a job. While every day new exotic plants (literally "strangers") streamed in from the shores of the Americas, Africa, and Asia, all too few local plants remained unknown; they had all been found.[66] In short, by 1727, the local flora had been so thoroughly institutionalized in Altdorf that its repeated writing and rewriting, Baier argued, was no longer necessary. The local countryside of Altdorf – or at least its plant species – were now "known." Nothing more remained to be done.

Indeed, by the early eighteenth century, many other towns across Europe were beginning to find themselves approaching the same position. While by far the greatest number of early local floras had been produced in the Holy Roman Empire, where confessional divides and decentralization had led to an especially high number of universities (and medical schools) competing for students, botanists in other areas of Europe had begun to follow the model of tiny Altdorf, and to produce their own local floras. In England, for example, the very first publication by the later-renowned botanist John Ray – in 1660, almost a half-century after Jungermann's & Hofmann's initial flora of Altdorf – was a local flora of the English university town of Cambridge, the first such work to be produced in the British Isles.[67] Within

---

[64] In effect, the botanical excursion might be seen as a form of "ceremony of possession" on a local level; see Patricia Seed, *Ceremonies of Possession: Europe's Conquest of the New World, 1492–1640* (Cambridge: Cambridge University Press, 1996) and Edward Muir, *Ritual in Early Modern Europe*, 2nd ed. (Cambridge: Cambridge University Press, 2005).

[65] Baier, *Horti Medici Acad. Altorf. historia curiose conquisita*, sig.)(4v.

[66] Baier, *Horti Medici Acad. Altorf. historia curiose conquisita*, sig.)(4v and)(5r.

[67] John Ray, *Catalogus plantarum circa Cantabrigiam nascentium* (Cambridge: excudebat J. Field, 1660).

another half-century, Cambridge itself, like Altdorf, had become the subject of numerous attempts to discover and itemize its plant life in its totality. Gradually, botanists in still other European countries adopted Jungermann and Hofmann's model as well.[68] Already in 1696, the wealthy naturalist Hans Sloane, later to be knighted and to donate his collections to the founding of the British Museum, found himself deciding to write up the results of a natural–historical journey to Jamaica not only in the form of a travelogue in English, but also as a local flora.[69] For, after all, only the local flora, written in Latin, could be read by scholars across Europe, unlike the travelogue, which because of its English words could be consumed only by a much more limited – dare we say "local"? – readership. And only a local flora could show that Sloane had made the most serious attempt possible to take inventory of the island's plant world, supplying both the names and the places that would show that he had married his experience to deep botanical learning. For botanists, Jamaica was now on the map, with many other places to follow.

TEXTS

The has focused on the *process* of making the local flora, as it unfolded in one especially revealing site; this section will examine the local flora as a *product*, that is to say, as a material and cultural artifact simultaneously. From its paper and binding to its deployment of Latin and vernacular words, the local flora was a concrete instrument constructed to achieve a distinct set of purposes. Long before Linnaeus, it evolved into a standardized and codified model for conveying information about local natural worlds, regardless of confession, language, or ethnicity – one that could be used in Catholic Europe

---

[68] In Sweden, for example, see Olaus Bromel, *Chloris Gothica, seu Catalogus stirpium circa Gothoburgum nascentium* (Göteborg: excudebat J. Rahm, 1694), and in France, Pierre Magnol, *Botanicum Monspeliense, sive plantarum circa Monspelium nascentium index* (Lyon: Ex officina Francisci Carteron, 1676) and eventually the famous systematist Joseph Pitton de Tournefort, *Histoire des plantes qui naissent aux environs de Paris* (Paris: De l'Imprimerie Royale, 1698). In Denmark, on the other hand, for many years authors seem to have preferred to label their works as encompassing the entire country: Simon Paulli, *Flora danica; det er, Dansk urtebog* (Copenhagen: Aff Melchiore Martzen, 1648); Peder Kylling, *Viridarium Danicum: sive catalogus plantarum indigenarum in Dania*, (Copenhagen: n.n., 1688). For a similar example in England, see William How, *Phytologia britannica, natales exhibens indigenarum stirpium sponte emergentium* (London: Typis Ric. Cotes, 1650), essentially a recompilation of information from Thomas Johnson's botanical excursions, discussed above.

[69] Hans Sloane, *Catalogus plantarum quae in insula Jamaica sponte proveniunt, vel vulgò coluntur, cum earundem synonymis & locis natalibus* (London: impensis D. Brown, 1696); Hans Sloane, *A Voyage To the islands Madera, Barbados, Nieves, S. Christophers and Jamaica, with the Natural History of the Herbs and Trees, Four-footed Beasts, Fishes, Birds, Insects, Reptiles, &c. of the last of those Islands* (London: printed by B. M. for the author, 1707–25).

as well as Protestant, indeed anywhere.[70] This model not only took hold throughout Europe, but also came to be exported to the rest of the world. Examining the local flora as an artifact – a carefully-constructed physical object containing some kinds of information but not others, and presenting this information in a particular way – reveals how the local flora came to embody a very specific set of concerns: not only those of the town, university, and garden, but simultaneously those of the new sciences. The local flora was more than a list of plants; it was quite literally a "book of nature." In the materiality of what it communicated, as well as what it did not, it not only represented local nature, but came for early modern Europeans to forge a new way of looking at it.

One feature of local floras which immediately strikes the modern reader is how physically small most of them actually were. Local floras tended to be tiny, belying the enormous quantities of information contained within them. Usually printed in more modest octavo or even duodecimo editions – in contrast to the lavishly illustrated herbals and florilegia of the time, which were often splendidly printed in folio – local floras were, in fact, designed to be small. They were "pocket" books in a literal sense, meant to be carried along on botanical excursions by medical students and their professors. Such portable textbooks had to be lightweight; and this is indeed how writers of local floras described their works. In self-deprecating fashion, they frequently referred to them in the diminutive, as "little books" or "little paper gifts."[71] The local flora was usually labeled an "opusculum," or "little work," rarely an "opus," no matter how much work might have gone into its making.[72] Though such self-deprecating language was far from unknown

---

[70] While the example discussed in the previous section, that of Altdorf, is a Protestant one, local floras also came to be produced at Catholic universities. One of the very earliest local floras, for example, was published at the Jesuit institution of Ingolstadt: Albert Menzel, *Synonyma plantarum, seu simplicium, ut vocant, circa Ingolstadium sponte nascentium* (Ingolstadt: typis Ederianis, per Elisabetham Angermariam, viduam, 1618). This would be true of other genres of local natural history more generally as well; for example, one of the first regional mineralogies was also produced under Catholic auspices: Friedrich Lachmund, *Oryktographia Hildesheimensis, sive admirandorum fossilium, quae In tractu Hildesheimensi reperiuntur* (Hildesheim: Sumptibus autoris, Typis viduae Jacobi Mülleri, 1669).

[71] See for example Nicolaus Oelhafen, *Elenchus plantarum circa nobile Borussorum Dantiscum suâ sponte nascentium* (Danzig: typis & impensis Georgi Rheti, 1643), 4; Giuseppe Monti, *Catalogi stirpium agri Bononiensis prodromus* (Bologna: apud Constantinum Pisarri, 1719), sig. [ii]v; and John Martyn, *Methodus plantarum circa Cantabrigiam nascentium* (London: ex officina Richardi Reily, 1727), v. On the practical, not merely metaphorical, use of books as gifts, see Natalie Zemon Davis, "Beyond the Market: Books as Gifts in Sixteenth-Century France," *Transactions of the Royal Historical Society*, 33 (1983): 69–88. For another example of an early modern "pocket" book, see Marcus Hellyer, "The Pocket Museum: Edward Lhwyd's *Lithophylacium*," *Archives of Natural History* 23 (1996): 43–60.

[72] Magnol, sig. [vi]v; Christoph Knauth, *Enumeratio plantarum circa Halam Saxonum* (Leipzig: sumpt. haered. Fried. Lanckisii, 1687), sig.)o(3r.

in the world of early modern publishing, where authors frequently sold their efforts short in order to emphasize all the more their patrons' status, in the case of local floras this language can, in fact, be taken in some ways quite literally. Extremely compact and condensed documents, local floras *were* indeed among the most dramatically minuscule and lightweight books available in early modern Europe.

While concurring on the question of their books' size, not all local florists, though, agreed on how to label them. This can scarcely be seen as a surprise, given that, as we have already seen, the term "local" was not yet in use in any of the languages involved, while the term "flora" might refer as easily to a goddess, or to the study of botany in general, as to a work of local natural history. Any given local florist might thus use a series of terms, seemingly interchangeably, to describe what he had written: most commonly, he might call it a "catalogue," an "index," and/or an "enumeration."[73] Each of these terms expressed, in a different way, the reliance of the local flora on the genre of the list. Whereas lists have traditionally been viewed as one of the simplest forms of writing imaginable, recent work has come to suggest that the list, in fact, may embody considerable complexity, and that its use is never "self-evident"; in short, that a set of choices always lie beneath the decision to cast a given set of information in the form of a list, rather than in prose, poetry, or other forms of communication.[74] The local flora, furthermore, used a specific kind of list. Whereas other natural–historical works might be allowed the luxury of "method," that is to say of a systematic organization by category, local floras were most often consigned to strict alphabetical order, in the fashion of the Renaissance "index."[75] Some of the earliest local plant catalogues were not even granted the minimal coherence of presentation

[73] These terms seem to have been used interchangeably; see, for example, Johann Jakob Dillenius, *Catalogus plantarum sponte circa Gissam nascentium* (Frankfurt: apud J. Maximilianum à Sande, 1719), sig. *3r, where the terms "index" and "catalogue" are presented as if synonymous; likewise, Knauth, sig.):(2r describes his book as a "catalogue," while its title presents it as an "enumeration." An early example, relatively unusual in that its introduction was written in the vernacular (not Latin), describes itself on its title page as a list and index both: Johann Schöpf, *Ulmischer Paradiss Garten, das ist: Eine Ferzeichnuss und Register, der Simplicien an der Zahl uber die 600. welche inn Gärten unnd nechstern Bezirck umb dess Statt Ulm zufinden* (Ulm: durch Johann Medern, 1622)

[74] Jack Goody, "The Recipe, the Prescription, and the Experiment," in *The Domestication of the Savage Mind* (Cambridge: Cambridge University Press, 1977), 129–145.

[75] See Dillenius, sig. [*6]r, and John Blackstone, *Fasciculus plantarum circa Harefield sponte nascentium* (London: typis H. Woodfall, junioris, 1737), v–vi. On the use of alphabetical order in the Renaissance, see Ann Blair, *The Theater of Nature: Jean Bodin and Renaissance Science* (Princeton, NJ: Princeton University Press, 1997), 66 and 162–163. In the early eighteenth century, as proponents of the classificatory systems of Ray, Tournefort and Rivinus began to clash with each other, some floras came to be arranged in "systematic" order instead; see for example Johann Caspar Gemeinhardt, *Catalogus plantarum circa Laubam nascentium* (Bautzen: apud Davidem Richterum, 1725), sig. A5r-A5v and 1–2.

as a coherent genre, like that of the "index"; instead, they were merely labeled as books of "synonyms," restricted to lists of plant names alone.[76] As "enumerations" of plants, the local floras that succeeded them were not designed to take up excessive space, whether in the pocket or on the printed page; instead, their task was to fit entire landscapes of botanical diversity within the covers of a tiny book, and in so doing within the confines of the format of the list.[77]

Local floras did not so much serve as substitutes for, but rather complemented, the enormous herbals and botanical reference works of the day, generally heavy volumes which no one could have actually carried on a lengthy excursion. The prior existence of these other works, to be consulted at leisure in the library, freed the authors of local floras from the burden of having to replicate their contents, in particular the lengthy entries herbals and similar works provided for each plant, containing elaborate descriptions of plants' physical appearance, popular folklore concerning them, their symbolism and emblematics, their common uses, and so forth. This enabled authors of local floras to concentrate instead on the new task they defined for themselves: providing stripped-down lists of nothing but mere names and places. Readers of local floras were thus referred to these other works, if they wanted fuller information on a given plant or plants; in his local flora of Cambridge, for example, English botanist John Ray took care to tell his readers about the four encyclopedic herbals that he was relying on.[78] Alternatively, a purchaser might order the bookbinder to interleave blank pages with the printed ones, as was commonly done for the purpose of individual note-taking.[79] This was necessary because most local floras were so economical with the printed page that they had virtually no margins; nonetheless, some users found the room to scrawl their notes *between* the lines. Local floras, then, were *not* intended to be all-purpose reference works, but rather primarily to serve a more specific set of uses and users.

---

[76] See, for example, the titles of Caspar Pilleter, *Plantarum tum patriarum, tum exoticarum, in Walachri, Zeelandiae insula, nascentium synonymia* (Middelburg: excudebat Richardus Schilders, 1610) and Menzel, *Synonyma plantarum.*

[77] Local floras often included some garden plants as well, on the logic that they had become naturalized in the place in question; see for example Johann Christian Buxbaum, *Enumeratio plantarum accuratior in agro Hallensi locisque vicinis crescentium* (Halle: In officina Libraria Rengeriana, 1721), sig.)( )( )( )(4v [sic]. In some cases, where local floras had been bound together with garden catalogues, or where the compiler had made the discussion to treat "not only the indigenous, but also the exotic" in a single volume (see Chapter 1 for a discussion of this phenomenon), even more of the world's botanical landscape could effectively be crammed into a student's pocket.

[78] Ray, sig. [*7]r.

[79] Reeds, *Botany in Medieval and Renaissance Universities*, 125. Many of the copies examined during the research for this book were bound in this way. On this practice more generally, see Ann Blair, "Note Taking as an Art of Transmission," *Critical Inquiry* 31 (2004): 85–107.

These users were, as has already been mentioned, primarily medical students. The didactic and pedagogical context of the local flora appears clearly in the tiny prefaces to be found at the beginning of most local floras, where "studious youth" were the ones almost always addressed. There is considerable evidence that the main audience for most local floras was medical students in particular. The title pages of many local floras, for example, specified that they were designed "for the use of the school of medicine" of the respective university, and as such, they were frequently issued by an "Academiae Typographus," or university printer.[80] Authors stressed the importance of an understanding of local plants for medical knowledge in general, and in particular for the education of those members of "most noble youth" (literally "adolescence") intending to be physicians.[81] Reinforcing their point with laudatory quotations from Galen, Hippocrates, and other authorities, urging the study of "simples" as essential for prospective doctors, they presented the local flora as a learning tool towards this aim. Throughout the seventeenth century, and on into the eighteenth, authors like Charles Deering in the English town of Nottingham highlighted Galen's opinion that "the doctor ought to have experience of all plants, if possible," and insisted on the need for medical students to acquire knowledge not only of the human body, but of living plants themselves.[82]

Local floras were designed to enable medical students to acquire a working knowledge of and ability to recognize those plants they would most commonly encounter as remedies in their future careers as physicians. Another practical goal lay behind this, that of enabling medical students to assume a set of duties they would be likely to encounter if awarded posts as town physicians, namely the responsibility of supervising town apothecaries and

[80] See for example the subtitle of Caspar Bauhin, *Catalogus plantarum circa Basileam sponte nascentium cum earundem Synonymiis & locis in quibus reperiuntur: in usum Scholae Medicae, quae Basileae est* (Basel: typis Johan. Jacobi Genathii, 1622). On these academic printers, precursors in many ways to the university presses of today, see Hilde de Ridder-Symoens, ed., 175 and 202–204. It was extremely rare for local floras to be published in anything other than a medical context; for an exception, though, see Johann Christoph Becmann, "Catalogus plantarum in tractu Francofurtano sponte nascentium," in *Memoranda Francofurtana* (Frankfurt an der Oder: prelo Friderici Eichornii, 1676), 72–80, where the local flora makes up a small part of a larger antiquarian miscellany on the history of the university.

[81] Christian Mentzel, *Centuria plantarum circa nobile Gedanum sponte nascentium* (Danzig: typis Andreae Hünefeldii, 1650), reprinted in Gottfried Reyger, *Tentamen florae gedanensis methodo sexuali adcommodatae* (Danzig: apud Daniel Ludwig Wedel, 1764–1766), v. 2, 206. On medical motivations more generally, see also Cornut, sig. [a ii]v; Magnol, sig. iiiʳ; and Giovanni Giacomo Roggieri, *Catalogo delle piante native del suolo Romano, co' loco principali Sinonimi, e luoghi natali* (Rome: per Felipe Cesaretti, 1677), 507.

[82] See for example the title-pages of both Bauhin and of Charles Deering, *Catalogus stirpium &c. or, A Catalogue of Plants Naturally growing and commonly cultivated in divers parts of England, more especially about Nottingham* (Nottingham: printed for the author by G. Ayscough, 1738).

ensuring that local plants were not used as cheaper substitutes for more expensive exotic ingredients.[83] But there was also an entire philosophy of medical education which went into the demand that students learn by "*autopsia*," a Greek term that had come to refer not only to the dissection or "autopsy" of cadavers, but also more literally to the act of "seeing for one-self."[84] Early modern physicians increasingly felt that it was necessary and beneficial for a prospective member of their number to have had the experi-ence of walking around, and observing the plants growing in their "natural" places. (One local florist opined, making a pun out of the similarity of the Greek word for "plants" and the Latin word for "soles," that the prospective physician had better have sound feet as well as a sound mind.)[85] Seeing plants in their places of origin helped one understand their "nature" more generally. Here the authors of local floras again cited Hippocrates and Galen, in order to argue that a knowledge of the natural world was necessary for the physi-cian in his broader role, as observer of "*physis*" (the original Greek word for nature). In this formulation, little distinction was made between that human nature seen in the body of the patient, and the external world of natural phenomena; both were equally the physician's province, as can be seen from the incredible array of treatises on all aspects of nature that streamed from early modern physicians' pens.[86] Local floras, then, incorporated an exten-sive set of ideals about the abilities and aspirations of future physicians.

In the opinion of the compilers of these books, they were sorely needed. Caspar Pilleter, an early Dutch local florist, reported that he had written his book so that students would not wander in a "labyrinth."[87] Others likewise saw the study of the natural world as a labyrinth, capable of confusing even their best and brightest students. Nature was certainly not a wilderness for them, and there is very little in local floras of that fear of the "wild" some historians have found in the writings of the Middle Ages.[88] But local

---

[83] Pilleter, sig. *4v; Johann Dietrich Leopold, *Deliciae silvestres florae Ulmensis* (Ulm: verlegts Johann Conrad Wohler, 1728), sig.)(4v and)(5r.

[84] Paul Ammann, *Supellex botanica* (Leipzig: sumptibus Joh. Christ. Tarnovii, 1675), sig.)(3r,)(4r,)(4v, and)(5v.

[85] Dillenius, sig. *5r.

[86] On physicians' broader claims to the knowledge of nature, or "physick," see Jerome J. Bylebyl, "The Medical Meaning of *Physica*," *Osiris* 6 (1990): 16–41; and Harold J. Cook, "Physick and Natural History in Seventeenth-Century England," in *Revolution and Continuity: Essays in the History and Philosophy of Early Modern Science*, eds. Peter Barker and Roger Ariew (Washington, DC: Catholic University of America Press, 1991), 63–80. See also Andrew Wear, "Making Sense of Health and the Environment in Early Modern Eng-land," in *Medicine and Society: Historical Essays*, ed. Andrew Wear (Cambridge: Cambridge University Press, 1992), 119–147.

[87] Pilleter, sig. [*6]v. See also Dillenius, sig. *5r and Heinrich Julius Meyenberg, *Flora Einbec-censis* (Göttingen: literis Josquini Woyken, 1712), 6.

[88] See for example Vito Fumagalli, *Landscapes of Fear: Perceptions of Nature and the City in the Middle Ages*, translated by Shayne Mitchell (London: Polity Press, 1993).

environments were clearly seen as territory unfamiliar, in many ways, to urban students headed outside city gates on botanical excursions; perhaps as unfamiliar as the plant kingdom (*"regnum vegetabile"*) itself. Students thus needed a "guide" to the natural world in both of these manifestations, and the authors of local floras saw themselves as providing the ideal guides, introducing the student to the world of botany and the world of the local countryside simultaneously. Just as students depended on *human* guides, namely their medical professors, to conduct them through the fields and forests that surrounded the *collegium* and to explicate the variety of natural forms they encountered there, so too did they need these carefully-formatted paper tools to guide them through the thought processes and experiences that they, as future physicians, were expected to acquire.

Local floras may not have been aimed *solely* at medical students, however, even though they may have been the primary targets. In the dedications composed for these works may be seen hints of another potential audience. Local floras were frequently dedicated to town or university officials, whether members of town councils, or deans, rectors, or any of the other myriad administrators and functionaries of the early modern university.[89] In this way, they show signs of being directed not only toward medical students, but also potentially toward these specific local audiences as well. For, as compilers suggested, a local flora could be every bit as much an "ornament" to a town or university as a recently-published history of a town, or even a newly-constructed building.[90] Amidst the culture of localism that prevailed in most early modern small towns – what has variously been called *campanilismo*, or the focus on the *Blick vom Kirchturm* (the view from the church tower) – local floras performed a variety of rhetorical functions that local functionaries would certainly have approved of. They presented a favorable image of the town or university to any outsiders who might learn of their publication. They testified to the skill of the scholars who had compiled them in the first place. In their devout praises of God as the creator of all natural things, they confirmed the piety of all involved. And, like the newly constructed botanical gardens, they performed the all-important task of commemoration – of individuals and of groups as well as of the landscapes, natural and social, within which they lived.

The ways in which local floras used language help show their multiple audiences and commitments. Local floras were commonly bilingual (or even trilingual), bridging the gap between the Latin of university life and the

---

[89] Examples are too numerous to mention, but the format was most commonly that of a long list of individuals (usually at least half a dozen), together with notes on their service to university or town.

[90] Thomas Bender, ed., *The University and the City: From Medieval Origins to the Present* (Oxford: Oxford University Press, 1988); Erich Maschke and Jürgen Sydow, eds., *Stadt und Universität im Mittelalter und in der frühen Neuzeit* (Sigmaringen: Thorbecke, 1977).

vernacular (or vernaculars) of everyday town life.[91] For example, a local flora published in England would usually combine Latin and English, while one from a German-controlled town in present-day Eastern Europe would be published in a combination of Latin, German, and the relevant Slavic language.[92] In this way, even within their tiny frames, local floras brought together multiple worlds, Latin and vernacular alike. This was reflected, furthermore, in the structure of the entries themselves. As we have seen, entries on individual plants were usually confined to the purpose of stating their nomenclature (i.e. "names") and location ("places"). Here there emerged a linguistic divide: whereas plants' various descriptive names were usually given in Latin, it was the vernacular that was most commonly used to describe the assorted places where plants might be found. In this way, local floras effectively identified Latin with the world of naming and scholarly activity, and the vernacular with the description of environments and local places.[93] In the multilingualism of the local flora, we see efforts to join together the universalism of scholarly description with a recognition that to truly capture the local, one had to use local language to do so.

This brings us to the final and most striking aspect of the local flora: what was *not* there. The local flora that arose in early modern Europe was a volume drastically pared down from the massive tomes of the Renaissance. Gone were many of the aspects of plants that had been discussed in herbals and similarly encyclopedic works: their visual appearance, uses, cultural meanings in myth, fable, and emblem, and so forth, as well as discussions of what local informants had said about them. Local floras were reduced to mere lists, "bare enumerations" of names and places. The omission of any

[91] On the role of Latin *vis-à-vis* the vernacular, see Peter Burke, "*Heu domine, adsunt Turcae*: A Sketch for a Social History of Post-Medieval Latin," in *Language, Self, and Society: A Social History of Language*, ed. Peter Burke and Roy Porter (Cambridge: Polity Press, 1991), 23–50; Ann Blair, "La persistance du latin comme langue de science à la fin de la Renaissance," in *Sciences et langues en Europe*, ed. Roger Chartier and Pietro Corsi (Paris: École des Hautes Études en Sciences Sociales, 1996), 21–42; Françoise Waquet, *Latin, or the Empire of a Sign*, translated by John Howe (London: Verso, 2001); and the classic article by Walter Ong, "Latin Language Study as a Renaissance Puberty Rite," *Studies in Philology* 56 (1959): 103–124. On botanical Latin in particular, see William T. Stearn, *Botanical Latin*, 4th ed. (London: David & Charles, 1992).

[92] Francke, for example, included words in the local Slavic dialect of Wendish, now called Sorb. Likewise, a local flora published in a town located in what is now Finland, parts of which were then under Swedish control, used a mixture of Latin, Swedish, and Finnish itself; see Elias Tillandz, *Catalogus plantarum quae prope Aboam tam in excultis, quam incultis locis huc usque inventae sunt* (Åbo [Turku]: excusus a Johanne L. Wallio, 1683). On this work, as well as a comparison of it with other local floras at the time, see Kurt Wein, *Elias Tillandz's 'Catalogus plantarum' (1683) in Lichte seiner Zeit erklärt und gewürdigt: ein Beitrag zur Geschichte der Botanik in Finnland* (Helsinki: Finnische Literatur-Gesellschaft, 1930).

[93] For a fuller discussion of these issues, see Alix Cooper, "Latin Words, Vernacular Worlds: Language, Nature, and the 'Indigenous' in Early Modern Europe," *Journal of East Asian Science, Technology, and Medicine*, forthcoming.

mention of the potential *uses,* whether medicinal or other, of the plants in question is especially significant here. Physicians had long been suspicious of herbals, especially vernacular ones, that had included any information as to use, on the grounds that this might lead to "misuse" and "abuse" on the part of laypeople attempting to treat themselves (and do without a doctor's services in the process). This would no longer be possible with local floras; their cryptically brief and codified contents effectively sealed them off from all but the most learned of those who consulted them outside of the approved pedagogical context, in other words, *without* a medical professor present to perform his task of "demonstrating" their contents. Even illustrations disappeared from the picture; very few local floras, especially given their tiny size, were to be accompanied by engravings, even woodcuts. All that remained were the cryptic bilingual entries for each plant, shorn of any other sets of association or connotation.

In this massive reduction of information about individual plants, down to a minimal level, combined with the great increase in the number of species listed, the local flora may thus be seen as taking part in a radical simplification of natural history. Historians of botany (and even those well outside the field) have repeatedly commented on the striking trend, in much late sixteenth-century and early seventeenth-century writing in natural history towards a new reductionist focus solely on the external visual *appearance* or morphology of organisms, with all their other attributes increasingly dropped from consideration.[94] In its own tendency towards radical simplification, the local flora can indeed be seen as following a similar pattern – but with one crucial difference. For in the local flora, even discussion of the *external appearance* of plants, the focus of so many other kinds of botanical writing, was dropped. The local flora, then, came to embody a drastic form of reductionism, one which reduced entire landscapes to mere lists.

PLACES

Yet local floras were not, in fact, entirely "bare enumerations"; nor can the thesis of radical simplification be accepted entirely unchallenged. For despite the removal of cultural, mythological, culinary, and even morphological description in the local flora, one category of information did remain, to assume a new importance: that about location. For it was *places* that came to be most thoroughly described in the local flora: the places where plants were to be found, their *loca natalia* or, literally, "birthplaces." This concern with location can be seen to have originated in the note-taking of Renaissance

---

[94] Michel Foucault, *The Order of Things: An Archaeology of the Human Sciences* (New York: Vintage, 1973); William B. Ashworth, Jr., "Natural History and the Emblematic World View," in *Reappraisals of the Scientific Revolution,* ed. David C. Lindberg and Robert S. Westman (Cambridge: Cambridge University Press, 1990), 303–332; Ogilvie, 203–206.

naturalists, who had increasingly, over the course of the sixteenth century, made it their practice always to jot down the location of any site where they had noted an interesting plant.[95] But the issue of location and origins was to become an even more prominent theme within works of local natural history, and local floras in particular. While recording the presence of individual plants, local floras linked them to the early modern landscapes and countrysides where they thrived. These landscapes were rarely presented as "wild," though, but rather as firmly linked to the community, whether that of the university or of the town. By reporting where plants were to be found, the local flora incorporated them into the institutions of a human world.

The local flora provides a unique window onto perceptions of place in early modern science, for its bulk consisted, in fact, of descriptions of place. The very structure of typical entries of plants in a local flora demonstrates this new emphasis. After providing the most authoritative names available for a given plant, and citing the references from which these names had been drawn, an entry proceeded immediately to the description of *where* it was that the plant could be found; this formed the remainder, and thus the bulk, of the entry. These discussions of location were often quite general, reporting the different kinds of habitats in which a plant was most commonly seen to thrive: whether in fields or in swamps, alongside the road, deep within forests, in hedges, in mountain pastures, and so forth. "Fields," "meadows," and "swamps" and their like were the "common places" of the local flora, just as humanist discourse developed its own set of "commonplaces" or *topoi*. Frequently more specific locations would also be given; for example, a particular mountain or road might be named, or even a specific house near which the plant might be found growing. John Ray, for example, mentioned in his local flora of Cambridge that plants could be found growing not only "in many places down the river Cam," but also "not far from Howes bowling-green," "at the Spittle-house," and even "on Jesus Colledge wall."[96]

Descriptions were thus often rich in detail. To the degree that they mentioned particular phenomena, to be found only in particular places, their aims may be seen as resonating with those of a culture newly focused on "facts" and particularity.[97] Yet the information about place they provided

[95] Ogilvie, 143–149. This emphasis on noting down details of locations seems to have culminated in the writings of the late sixteenth-century botanical traveler Carolus Clusius: see for example his *Rariorum aliquot stirpium per Hispanias observatarum historia* (Antwerp: Ex officina Christophori Plantini, 1576) and his *Rariorum aliquot stirpium per Pannoniam, Austriam, et vicinas...historia* (Antwerp: Ex officina Christophori Plantini, 1583).

[96] Ray, 4, 13, 14, 6. A manuscript local flora from the mid-eighteenth century (compiler unknown), titled "Catalogus plantarum in Praefectura Wildershausen et proxime adjacentibus sponte nascentium," even included weeds growing "wild in my own garden": Universitätsbibliothek Göttingen, Hist nat 96, fol. 25.

[97] Lorraine Daston, "Baconian Facts, Academic Civility, and the Prehistory of Objectivity," *Annals of Scholarship* 8 (1992): 337–363.

was specific only to a certain extent. The kind of detail to be found in experimental reports of "matters of fact," for example the precise citing of individuals and dates, did not appear, nor may any forms of numerical precision be found. The local flora, then, expressed a different kind of particularity than that to be found in the laboratories of the new scientific societies. Plants were framed instead as particular in the ways in which they could be "placed" and located within specific early modern landscapes.

Plants were presented as living objects, not fixtures. Entries typically began with the word *crescit* ("it grows") or *habitat* ("it lives," whence our modern word "habitat," meaning environment). This interest in plants as living, growing beings was likewise reflected in the frequent references to plants as "growing of their own accord." Plants *grew*, and in their growth reflected the creative capacities of the (often female) nature that produced them.[98] Many local floras took special care to record not just the places in which plants grew but also the times at which they began to flower.[99] But plants' distinguishing characteristic among living beings was that they were literally rooted to the place where they grew. Unlike animals, which possessed means of locomotion, plants were confined to certain places, and were thus even further defined by these places.[100] It was, of course, recognized that owing to their airborne seeds, as well as transplantation by humans, many plants could "travel" quite effectively; indeed, over the course of the early modern period, many new plants from Africa, the Americas, and Asia gradually became naturalized in Europe, often as a result of specific efforts at plant introduction and acclimatization.[101] However, it was clear to all involved

[98] See for example the discussion of the role of Mother Nature in Johann Chemnitz, *Index plantarum circa Brunsvigam trium ferè milliarium circuitu nascentium* (Braunschweig: Typis & sumtibus Christophori Friderici Zilligeri, 1652), sig. [Ai]v. Another local florist referred to plants' locations as their "cradles" (*incunabula*), in language that may perhaps have drawn on this kind of association: see Johann Georg Duvernoy, *Designatio plantarum circa Tubingensem arcem florentium* (Tübingen: Typis & impensis Georgii Friderici Pelickii, 1722), sig. A3r.

[99] See Knauth; Menzel; Blackstone; and especially Theophilus Kentmann, *Tabula locum et tempus, quibus uberius plantae potissimum spontaneae vigent ac proveniunt, exprimens* (Wittenberg: apud Haeredes Samuelis Seelfisch, 1629), with its many subsequent editions.

[100] François Delaporte, *Nature's Second Kingdom: Explorations of Vegetality in the Eighteenth Century*, trans. Arthur Goldhammer (Cambridge, MA: The MIT Press, 1982), 149–185.

[101] Recent scholarship has done much to uncover eighteenth- and nineteenth-century naturalists' fascination with acclimatization projects: see for example Lisbet Koerner, *Linnaeus: Nature and Nation* (Cambridge, MA: Harvard University Press, 1999); E. C. Spary, *Utopia's Garden: French Natural History from Old Regime to Revolution* (Chicago, IL: University of Chicago Press, 2000); Harriet Ritvo, "At the Edge of the Garden: Nature and Domestication in Eighteenth- and Nineteenth-Century Britain," *Huntington Library Quarterly* 55, 3 (1992): 363–378; and Michael A. Osborne, "Acclimatizing the World: A History of the Paradigmatic Colonial Science," in *Nature and Empire: Science and the Colonial Enterprise*, ed. Roy MacLeod (Chicago, IL: University of Chicago Press, 2000), 135–151. However, many sixteenth- and seventeenth-century botanists, including many local florists, were also highly involved in efforts to get exotic species to grow in European gardens. See

that a wide range of limiting factors were involved, in particular climate and soil. Far too many plants introduced from far-away places perished during cold European winters, despite all the care lavished on them, for botanists not to insist on the significance of plants' attachment to their "native" climes. In the tradition of what has been called Renaissance environmentalism (that is to say, the belief that geography determined living beings' characteristics), plants were viewed as physically rooted in particular environments in a way animals never could be.[102]

This may be seen by examining further the concepts used to express these ideas in the local flora. The importance of knowing a plant's *locus natalis*, or "birthplace," for example, was stressed again and again. For the *locus natalis* of a plant was seen as reflecting a natural (and also divine) order, in which God had placed plants in their proper habitats.[103] Each plant was thus presented as possessing its own *patria*, or "native land" (literally, "fatherland"), a word loaded with human associations. In investigating the *patria* of a plant, then, local florists sought clues to its very identity. Rather than turning to a plant's origins in time, as evolutionary thinkers of the late eighteenth and nineteenth centuries were to do, early modern naturalists searched for plants' origins in space, that is to say, within a *geographical* rather than a chronological framework for understanding. By investigating the "native land" of a plant, they inquired into its very organic origins, as an organism that called a certain kind of place home.[104]

Local florists were fully aware that the "natal places" of plants were often very different from those of people. Many plants grew in marshes, for example, as well as in other places hardly viewed as desirable locations for human habitation. Local florists, however, repeatedly boasted of their willingness to seek out plants wherever they grew, regardless of personal convenience or the lack thereof. And in the course of their botanical excursions, they began to note some sites as of particular interest. Mineral springs, for example, often seemed to harbor rare plants; the early modern period saw considerable interest on the part of physicians in these places, and many local florists

Chandra Mukerji, *Territorial Ambitions and the Gardens of Versailles* (Cambridge: Cambridge University Press, 1997) for an example of a highly organized such effort; quite often, though, these kinds of projects were carried out simply through private correspondence networks.

[102] Clarence J. Glacken, *Traces on the Rhodian Shore: Nature and Culture in Western Thought from Ancient Times to the End of the Eighteenth Century* (Berkeley: University of California Press, 1967), 429–460.

[103] See for example Karl Schaeffer, *Deliciae botanicae Hallenses* (Halle: typis Christophori Salfeldii, 1662), sig.)(2r.

[104] In fact, if a botanical author omitted reference to a specific plant's *patria*, readers might become suspicious as to the validity of the information about that plant; see for example Johann Jakob Dillenius, *Hortus Elthamensis* (London: sumptibus auctoris, 1732), II, 403, where he blamed another author for not having found the "special place" of a given plant. Hoffmann, *Montis Mauriciani*, 30, on the other hand, praised the English botanist John Ray for having diligently noted the "natal places" of every plant he listed.

also authored separate accounts of local spas.[105] Mountains likewise often revealed themselves to be excellent places for finding not only unusual minerals, but also rare plants. It has become a truism that mountains were, in the early modern period, seen as wild and forbidding places, and that only towards the end of the eighteenth century did this negative stereotype begin to slip away, as they began to be viewed as potential sites for tourism.[106] But in fact early modern florists took quite avidly to the slopes, as is manifested in the frequency with which they found rare species there. Certainly this movement had a venerable humanist lineage, dating back to Renaissance physicians' expeditions to Monte Baldo, and even further back to Petrarch's earlier famous trip to Mt. Ventoux; but this prehistory does not prepare one for the degree to which early modern botanists embraced mountains as primary sites for their fieldwork. Caves were another kind of "special" place, one not part of the daily itinerary of most townsfolk or rural inhabitants, but nonetheless often harboring unusual species.[107] Medical professors prided themselves on their inclusion of such places in botanical excursions, and on the thoroughness with which they uncovered both new plants *and* new places to find them in.

Yet however far from the city the compiler of the local flora might ramble, ultimately he was bound by it. Early modern local floras were, quite literally, centered on the human world, in this case, on the town, with the university as the site of knowledge within it. This can be seen, for example, in the way in which the titles of many early modern local floras defined their contents in direct relation to the towns their compilers lived in. A title might thus specify that the plants included would be those within a radius of, say, three, four, or five miles from the town in question.[108] The town, then, representing the

---

[105] On the fascination of local springs for physicians, see Katharine Park, "Natural Particulars: Medical Epistemology, Practice, and the Literature of Healing Springs," in *Natural Particulars: Nature and the Disciplines in Renaissance Europe*, eds. Anthony Grafton and Nancy Siraisi (Cambridge, MA: The MIT Press, 1999), 347–367; and Roy Porter, ed., *The Medical History of Waters and Spas* (London: Wellcome Institute for the History of Medicine, 1990).

[106] See for example Marjorie Nicolson, *Mountain Gloom and Mountain Glory* (Ithaca, NY: Cornell University Press, 1959) and, for an excursion into the sources of these cultural stereotypes, Simon Schama, *Landscape and Memory* (New York: Alfred A. Knopf, 1995), 411–423.

[107] Oelhafen, column 80, noted for one kind of thistle that it was, in fact, only to be found in such "hidden places."

[108] See for example the title-page of Knauth, which listed a radius of three miles. Of course, the unit of the "mile" itself was one which could vary significantly from place to place, reflecting local variation in early modern systems of measurement. Caspar Bauhin's local flora of Basel was unusual in that it claimed to confine itself to the space of a mere mile (Bauhin, title-page); in fact, however, Bauhin, like other other compilers, did not adhere strictly to these limits, but included listings for numerous plants to be found on the so-called *Mons Wasserfall*, a popular destination for botanizing among Basel medical students.

Figure 7. Topographical Map of the Altdorf Vicinity. Altdorf lies at the center of this map, while the trading center of Nuremberg is on the top left. Note the concentric circles. From Moritz Hoffmann, *Florae Altdorfinae deliciae hortenses* (Altdorf: typis G. Hagen, [1660]). Library of the Arnold Arboretum, Harvard University, Cambridge, Massachusetts, USA.

human world, symbolically and practically anchored the local flora's investigations of its surrounding territory. A graphic representation of this may be seen in the topographical map which accompanied Moritz Hoffmann's *Florae Altdorfinae deliciae sylvestres*, showing the small town of Altdorf at the center of the map, with a series of concentric circles expanding outwards from it (see Figure 7). Here too the town appears as the focus of the naturalist's world, with the assorted symbols for trees, swamps, and mountains, which adorned the remainder of the map, presented through the device of the concentric circles as centered on the human institution of the town.

This is a recurring theme in the earliest generation of local natural histories. In many local floras nature appears as virtually integrated into the substance of the early modern town, in the form of listings of mosses and creepers to be found growing on the city walls themselves – right on the boundaries between inside and outside. As has already been mentioned, plants were also often listed as growing next to some prominent person's home, or in cultivated fields. Authors of local floras were very aware that much of the land they described was privately owned, or in other ways lay within human possession. In the spare, abstract list format of the local flora, almost all of the rich detail and description of the sixteenth century herbals had been dropped, *except* for these kinds of information about place and location.

Local natural histories thus represented perhaps the ultimate in a what might be called a "located knowledge." In their pages, local nature could not be detached from the world of the town, nor could this knowledge be utterly detached from the circumstances of its creation.

Yet it is crucial to recognize that local floras simultaneously reached in many ways beyond the boundaries they referred to. Compilers of local floras frequently compared the fertility and variety of the area surrounding their towns with those of *other* areas. They cited other local floras from different locales, and sent copies of their own to scholars scattered across Europe. Almost every compiler of a local flora included an extensive bibliography of this far-flung floristic literature.[109] Compilers seem, in fact, to have been quite aware of as well as keenly interested in other places. While engaging in the peripatetic culture of the early modern university, the vast majority of them had traveled, not only on botanical day-trips, but considerably further afield, whether on academic pilgrimage to the renowned universities of Italy, or to other burgeoning centers of learning in the Netherlands, France, Germany, and England.[110] Florists' peregrinations had often landed them far from their initial *patria*; and even as they took inventory of the botanical riches around their new homes, they compared and contrasted them with those of places elsewhere.

Even as the genre of the local flora emerged within a distinct set of European contexts, then, it must be seen as reaching well beyond any simple or uncomplicated transcription of a unitary "local" world. Compilers of local floras set forth in their cryptically brief and densely Latinate works a kind of natural knowledge very different than any that would have been provided by other townsfolk, let alone farmers residing outside town walls. In the process, they established their own forms of natural description as the benchmark for true knowledge of local natural worlds. In florists' correspondence with other scholars, they frequently gave as a reason for writing their own local floras, simply: "because they were lacking." Local floras, like botanical gardens, had become objects of prestige, emulation, and even competition between different sites within Europe. Talking about local nature thus provided a way of discussing and validating human identity as well, in particular the civic identity of the individuals who wrote local floras; but it also provided opportunities for looking *outwards* at the world, for engaging in a spirit of comparison and even of cosmopolitanism. While claiming to represent the local, local floras engaged by their very nature with the knowledge of a broader world.

---

[109] See for example Schaeffer, sig.)(2r through sig. [)(4]v, which provided what was, for a tiny preface to a tiny book, a voluminous listing of other local florists.

[110] Antoni Maczak, *Travel in Early Modern Europe*, translated by Ursula Phillips (Cambridge: Polity Press, 1995); and Justin Stagl, *A History of Curiosity: The Theory of Travel, 1550–1800* (Chur, Switzerland: Harwood Academic Publishers, 1995).

3

From Rocks to Riches: The Quest
for Natural Wealth

In the year 1545, Spanish explorers in Peru discovered tremendous quantities of silver ore at Potosí. Following Columbus's New World voyages in 1492 and his reports of the vast mineral treasures to be found in the Americas, the Spanish had established a number of smaller mines in their new colonies. But none of them matched Potosí, the silver output of which soon became legendary, a symbol of the wealth to be gained from colonial ventures. Fleet after fleet crossed the Atlantic, transporting the precious bullion to Spain, and leaving powerful effects on European economies in their collective wake, from a lasting inflation of prices to the faltering and eventual stagnation of the previously-famed German silver mines. Paradoxically enough, though, even as all eyes turned to Potosí and the other New World sources of precious metals that succeeded it, a number of Europeans began to call for the exploration of what they saw as Europe's own "mineral wealth." Lying buried beneath the surface of their own lands, they claimed, were subterranean treasures hitherto ignored. And precious metals formed only a small fraction of these. *All* areas possessed mineral wealth, they argued, that could be put to use, if only residents would recognize its true value.

Starting around the middle of the seventeenth century, a new genre of local natural history appeared in Europe, this time centered on the observation and description of the mineral kingdom. Following the conclusion of the Thirty Years War in 1648, during the lengthy period of reconstruction that followed, a series of works that we might now term "regional mineralogies" began to be published, outlining the mineral riches of various areas in Europe, particularly in the German territories.[1] These works bore strong links with the local flora tradition which they liberally cited; indeed, their authors were equally likely to be professors of medicine in small European university towns. Treatises on "subterranean Braunschweig," "underground

---

[1] Though *oryktographia* was the most common label for these works (from the Greek root *orykto-*, or rock), with *mineralogia* and *lithographia* somewhat less popular, I will refer to them as "regional mineralogies" for simplicity. However, it should be stressed that these works, in their early modern forms, did not share the modern geological subdiscipline of mineralogy's focus strictly on structure and chemical composition, but rather discussed a far broader range of rocks' attributes and uses.

Silesia," and so forth, however, helped to extend the reach of local natural
history beyond the plant kingdom into another kingdom of nature: that of
minerals, or, more accurately, "fossils" *(fossilia)*. Including all things "dug
up out of the earth," this term was used in the early modern period to encompass minerals, metals, earths, and "figured stones" as well as all other sorts
of rocks and stones, from crystals to landscape marble.[2] In the process of
describing these many and varied objects, a new model for doing local natural history arose, one which was to provide an alternative to the increasingly
formalized pattern of the local flora.

   This chapter will sketch some of the characteristics that this new form of
writing about natural history took, as writers of local floras expanded their
reach during the late seventeenth and early eighteenth centuries into the territory of the underground. One of the most prominent themes of the regional
mineralogy, as we shall see, was the concept of "natural wealth." Whereas
local floras had invoked the *fertility* of the regions they described, approvingly citing God's and nature's gifts of vegetation springing spontaneously
forth from a bountiful earth, regional mineralogies increasingly drew instead
on notions of *productivity*, on the necessity of human ingenuity, intervention, and labor to ensure the proper discovery and deployment of natural
riches. In the process of recounting the natural wealth of each territory,
regional mineralogies offered their readers not only inventories of natural
productions, but also tools for locating them and making them accessible,
for making nature's hidden treasuries reveal their true treasures. This shift
in language was reflected in a shift in authorship and audience. Authors of
regional mineralogies came to include not just university medical professors,
but court and town physicians, who appealed to broader audiences to realize
their visions of the advantages to be gained from the recognition of natural
riches.

   In this language of natural wealth, we can glimpse a vision of the world
still somewhat removed from that implied by the modern term "natural
resources," with its utilitarian connotations. For the early modern regional
mineralogy reveals quite different conceptions of what might be called
"value" or utility. Perceived economic utility alone did not make those natural objects described in regional mineralogies into "natural riches"; rather, a
much broader sense of utility was at work, one including religious, political,

---

[2] See Martin Rudwick, *The Meaning of Fossils: Episodes in the History of Palaeontology*,
2nd ed. (Chicago, IL: University of Chicago Press, 1976), 1–2. A number of recent works
have studied developments in mineralogy and related disciplines during this period, before
the advent of "geology": Rachel Laudan, *From Mineralogy to Geology: The Foundations
of a Science, 1650–1830* (Chicago, IL: University of Chicago Press, 1987); Gabriel Gohau,
Les sciences de la terre aux XVIIe et XVIIIe siècles: Naissance de la géologie (Paris: Albin
Michel, 1990); F. Ellenberger, *Histoire de la géologie, II: 1660–1810* (Paris: Lavoisier, 1994);
and Rhoda Rappaport, *When Geologists Were Historians, 1665–1750* (Ithaca, NY: Cornell
University Press, 1997).

intellectual, and above all symbolic dimensions. Local specimens showed their varied usefulness in proving the existence of divine Providence, highlighting princes' interest in the products of their lands, and displaying the infinite variety of the world. Furthermore, local specimens came to acquire considerable meaning because of their very localness. Naturalists came to pride themselves on their knowledge of, and ability to interpret, the natural productions of their own areas. Through their access to local specimens, naturalists residing in obscure corners of the Empire were able to connect themselves with an increasingly international community, as they came to offer their specimens and expertise to those residing in other areas. Contrasting the minerals of their districts with those to be found elsewhere, naturalists created a culture of comparison centered on the exchange and interpretation of local specimens.

MINERAL KINGDOMS

Authors of local floras had long expressed an interest in extending their descriptions of local natural worlds into regions beyond the purely botanical. In his early flora of the Harz forest, for example, Johann Thal had outlined a vision of a much more comprehensive work, which would necessarily include a discussion of the area's rivers, its fishes and quadrupeds, and last but not least, its "metals, fossils, and similar things."[3] Thal's vision was not to be fulfilled; nor was that of Johann Gottsched, a century later, who republished an earlier compiler's *Flora Prussica*, and promised his readers that he considered this work only the "first part of the natural history of Prussia," expressing his hopes of someday publishing "a second part about the animals of Prussia, and a third about its minerals."[4] Although none of these projects came to fruition, they indicate an interest among authors of local floras in exploring "natural history" beyond the confines of the vegetable kingdom. The study of plants, after all, as they repeatedly pointed out, was only one part of this broader enterprise. The mineral specimens that they found on their botanical excursions offered a logical next step.

A lengthy tradition of writing about the mineral world already existed, in the mining literature of Europe and especially of the German territories. Mining enterprises had flourished in many parts of the medieval Holy Roman Empire, far more than elsewhere in Europe, generating in the process a wide variety of documents ranging from lists of prices and procedures to

---

[3] Johann Thalius, *Sylva Hercynia, sive catalogus plantarum sponte nascentium in montibus & locis plerisque Hercyniae Sylvae* (Frankfurt: apud Iohannem Feyerabend, 1588), 3–4.

[4] Johann Gottsched, *Flora prussica, sive Plantae in regno Prussiae sponte nascentes* (Königsberg: sumptibus Typographiae Georgianae, 1703), sig.)o(2r; this work was a revised and expanded version of the long-out-of-print Johann Loesel, *Plantas in Borussia sponte nascentes* (Königsberg: typis Paschalii Mensenii, 1654).

descriptions of particular mines. The early sixteenth century saw a marked growth in this literature, as humanist scholars, often medically trained, came to involve themselves in the administration of mines, and to attempt to record their operations in encyclopedic detail.[5] Thus we see from the first half of the century, for example, a series of dialogues explaining the various kinds of knowledge that the miner had to possess, knowledge of such abstruse matters as the slope and dip of veins; and around the middle of the century various works intended to reproduce not only the theoretical knowledge of mining, but also its artisanal and mechanical foundations. Georg Agricola's famous *De re metallica* ended up providing the most thorough compendium yet, fully illustrated and based on his experiences in the mining towns of Saxony, of the full range of activities and techniques involved in the industry.[6]

Renaissance physicians had proved among the most avid collectors of *naturalia* of all types, geological specimens included. The famous Swiss polymath Conrad Gesner wrote a treatise on minerals based largely on the specimens in his own collection, as did the papal physician Michele Mercati in Rome, though his wasn't published till some time later.[7] Indeed, the cabinet of Johann Kentmann in the late sixteenth century provides one of the first examples of a mainly regional catalogue.[8] Physicians legitimated their mineralogical interests not so much on pedagogical grounds – after all, most rocks and fossils possessed much less obvious medicinal uses than plants – as on broader conceptions of their medical role. Drawing on sentiments articulated in both Hippocratic and Galenic authors, and revived by Paracelsus and his followers, they argued for the physician's role as an interpreter of natural matters in general, as an inquirer into "*physis.*"[9] With all of the natural world the physician's province, he was not limited to the investigation of

---

[5] Pamela O. Long, "The Openness of Knowledge: An Ideal and Its Context in 16th-Century Writing on Mining and Metallurgy," *Technology and Culture* 32 (1991): 318–355; Manfred Koch, *Geschichte und Entwicklung des bergmännischen Schrifttums* (Goslar: Hübener, 1963).

[6] Georg Agricola, *De re metallica libri XII* (Basel, 1556). Though Spanish America became a hub of metallurgical activity, especially around Potosí, little of the skilled knowledge involved was published until Alvaro Alonso Barba's *Arte de los Metales* (Madrid: en la Imprenta del Reyno, 1640).

[7] Conrad Gesner, *De omni rerum fossilium genere, gemmis, lapidibus, metallis, et huiusmodi* (Zürich: excudebat Iacobus Gesnerus, 1565); Michele Mercati, *Metallotheca. Opus posthumum, auctoritate, & munificentia Clementis undecimi pontificis maximi e tenebris in lucem eductum; opera autem, & studio Joannis Mariae Lancisii, archiatri pontificii* (Rome: ex officina Jo: Mariae Salvioni, 1717). For more on the latter work, see Alix Cooper, "The Museum and the Book: The *Metallotheca* and the History of an Encyclopaedic Natural History in Early Modern Italy," *Journal of the History of Collections* 7, 1 (1995): 1–23.

[8] Johannes Kentmann, *Nomenclaturae rerum fossilium, quae in Misnia praecipue, & in aliis quoque regionibus inveniuntur* (Zürich: excudebat Jacobus Gesnerus, 1565).

[9] Jerome J. Bylebyl, "The Medical Meaning of *Physica*," *Osiris* 6 (1990): 16–41 and Harold J. Cook, "Physick and Natural History in Seventeenth-Century England," in *Revolution and Continuity: Essays in the History and Philosophy of Early Modern Science*, ed. Peter Barker and Roger Ariew (Washington, DC: Catholic University of America Press, 1991), 63–80.

local plants, but was equally entitled to pursue the study of whatever other natural phenomena his path might cross.

The rise of the local flora as an accepted channel for medical activity seems to have offered a powerful incentive for the parallel study of minerals. Commencing in the second half of the seventeenth century, we see an increasing interest in mineral matters among physicians, often the very same ones who wrote local floras. Moritz Hoffmann, the Altdorf author who wrote such elegant accounts of his botanical excursions, is a case in point. In his 1694 narration of his trip up the Moritzberg, for example, discussed in the previous chapter, he had repeatedly mentioned interesting rock outcrops and other mineralogical sites.[10] But his interest in fossils can be detected much earlier, in his 1660 *Florae Altdorfinae deliciae hortenses* or garden catalogue, with its illustration of the garden displaying *fossil*, not plant, specimens neatly lined up in the fields outside the garden (see again Figure 6 in Chapter 2).[11] Apparently, plant-hunting excursions into the countryside might bring minerals, among other objects, to the physician's eye.

The authors of the earliest regional mineralogies, then, primarily physicians, repeatedly cited local floras as precursors to their efforts.[12] Praising the labors of their botanically-inclined predecessors in contributing to the "natural history" of the regions in which they lived, they set themselves in this same broader tradition of writing.[13] Johann Jakob Baier, for example, the author of an *Oryktographia Norica* (1708), which investigated the region around Nuremberg, cited Hoffman's *Florae Altdorfinae* as an inspiration; local floras had long examined those natural productions *above* ground and it would be well worth the effort, he felt, "if after having gotten to know and traversed the surface of the flourishing earth, if I could also show them [his students] the way to those inner things worth looking. . . . "[14] Like authors of local floras, authors of regional mineralogies insisted on the value of

---

[10] Moritz Hoffmann, *Montis Mauriciani in agro Leimburgensium, medio inter Norimbergam & Hirsbruccum* (Altdorf: typis Henrici Meyeri, 1694), 4, 9, 18, and 21.

[11] Moritz Hoffmann, *Florae Altdorfinae deliciae hortenses sive Catalogus plantarum horti medici* (Altdorf: typis G. Hagen, 1660).

[12] There were exceptions, i.e several authors of regional mineralogies were not trained in medicine, but rather natural philosophy or theology; see for example Johann Brunner, *Dissertatio physica de figuris variarum rerum in lapidibus & speciatim fossilibus comitatus Mansfeldici* (Leipzig: literis Joh. Georgi, 1675).

[13] Most did so by including the authors of local floras in their lists of works consulted, or by referring to them within the text as necessary, though footnotes were increasingly coming into use: for an extreme example, see Christian Gabriel Fischer, *Lapidum in agro Prussico sine praejudicio contemplandorum* (Königsberg: literis Reusnerianis, 1715). On the rise of the footnote at the time, see Anthony Grafton, *The Footnote: A Curious History* (Cambridge, MA: Harvard University Press, 1997).

[14] Johann Jakob Baier, *Oryktographia Norica, sive rerum fossilium et ad minerale regnum pertinentium, in territorio Norimbergensi eiusque vicinia observatarum succincta descriptio* (Nuremberg: impensis Wolfgangi Michahellis, 1708), sig.)(2r.

systematically traversing local fields and forests in search of the opportunity to see specimens for themselves (*autopsia*), in their natural places of origin.[15]

This new generation of authors, criticizing (somewhat unfairly) their humanist predecessors, insisted that it was not enough to consult previous writings on minerals; rather, it was necessary to inspect actual specimens, if possible not merely in boxes as part of a collection, but by taking actual trips to the sites where they were to be found, so that the rock formations could be examined and hypotheses as to their origins developed. Friedrich Lachmund, for example, the author of the early *Oryktographia Hildesheimensis* (1669), acknowledged his partial reliance on authors "from the previous century," such as Valerius Cordus and Georg Agricola himself; but he stressed that "in order to become more certain about these fossils, I myself diligently traversed and examined the mountains, the valleys, the quarries and the sandy places between the city and the Steurwald fortress several times, by longitude and latitude."[16] Lachmund, in short, took care to differentiate himself from previous authors, stressing that he had found not only the various kinds of rocks described by Agricola, but also "other very rare ones," including one which "as far as I know, has neither been described nor depicted by anyone."[17] His friend Friedrich von Hagen, in a letter included in the volume, referred to Lachmund's status as an "ocular witness," maintaining that it would be a "crime" if he and other such possessors of firsthand knowledge did not commit it to print.[18] In all these ways, the regional mineralogy showed itself to be an outgrowth of efforts similar to those which had originally generated the local flora.

Yet new possibilities and tensions arose in the writing of the mineral world. When compilers of local floras turned their attention to nearby "fossils," they faced numerous challenges. Whereas plants were usually reducible to species or recognizable natural kinds, mineral masses, often amorphous and mixed in composition, could rarely be categorized or written-up as easily as species of plants; they required fuller description. Another factor complicating writing about the mineral kingdom was the increasing attention that came to be

---

[15] See Baier, *Oryktographia Norica*, 35.

[16] Friedrich Lachmund, *Oryktographia Hildesheimensis, sive admirandorum fossilium, quae In tractu Hildesheimensi reperiuntur* (Hildesheim: sumptibus autoris, Typis viduae Jacobi Mülleri, 1669), sig. (o)2r (owing to a printing error this appears as (o)3r). Likewise, Gottlob Friedrich Mylius, author of a *Memorabilium Saxoniae subterraneae, i.e. Des Unterirdischen Sachsens seltsamer Wunder der Natur* (Leipzig: In Verlegung des Autoris, 1709), sig.):(v, complained that previous authors on the mineral world had not included specimens' place of origin (*den locum natalem*); his reversion into Latin here suggests a possible familiarity with discussions of the importance of the *locus natalis* in local floras and other botanical works (see Chapter 2).

[17] Lachmund, sig. (o)2r; see above re printing error. Johann Jakob Lerche likewise claimed in his *Dissertatio inauguralis physica-medica sistens Oryctographiam Halensem sive fossilium et mineralium in agro Halensi descriptionem* (Halle: Typis Joh. Christiani Hilligeri, 1730), 6, that many of the specimens he was about to describe were "still unknown."

[18] Lachmund, sig. (?)[3]v [sic].

paid during this period to "figured stones," many of which we would nowadays call fossils: rocks which seemed to bear images on them, whether of fish or plants, or even stranger creatures and objects. Historians of geology have traced the lengthy debates over the origin of figured stones, with the two main camps being those who thought that the images were created by Aristotelian "exhalations" or by a "plastic force" within the rock, and those who believed that they might be the actual organic remains, or impressions, created by long-ago creatures.[19] Though the latter view was increasing in popularity towards the end of the seventeenth century, debate was lively, and virtually all who wrote about rocks, including authors of regional mineralogies, expressed some opinion. With the infinitely diverse images that sometimes appeared impressed upon figured stones, however, it was difficult to classify them into species (although some efforts in this direction were made). These rocks were often genuinely one-of-a-kind, possessing individuality in a way plants could not. Hence simple lists rarely sufficed to convey any sense of adequate description, as in the local flora. Illustrations, for example, might be necessary to convey the appearance of an especially singular or extraordinary specimen; and such visual aids, which had been relatively rare in the local flora, did indeed make a reappearance in the regional mineralogy. Detailed descriptions of location also assumed increasing importance; it might make a great deal of difference, for instance, whether a marine fossil was found on a swampy lowland, or on a mountaintop high above sea level, and it might also make a difference what other rock formations accompanied it. Numerous other sorts of information about form, appearance, and location likewise came to play a key role in the regional mineralogy.

In this way, authors of regional mineralogies came to adopt a much fuller form of description than the local flora, with the latter's reduction to the format of the list. Writing far more expansively, they included lengthy discussions not only of physical appearance and geographical provenance, but also of local variation, folklore, and popular use. In this way, the regional mineralogy might be seen as having reinstated the humanist style of extensive description, which had been stripped from the local flora. But the form the regional mineralogy took was in fact very different from that of many of its humanist predecessors, such as the works of Aldrovandi; such categories as "hieroglyphics," "symbols," and "fables" found short shrift in the regional mineralogy, giving way to emphases on individuality and place.[20] Though written by the same medical professors who had stripped the local flora down to its bare essentials, the regional mineralogy thus presents to the

[19] Rudwick, 1–45; on the former explanation, see also Paula Findlen, "Jokes of Nature and Jokes of Knowledge: The Playfulness of Scientific Discourse in Early Modern Europe," *Renaissance Quarterly* 42 (1989): 292–331.
[20] William B. Ashworth, Jr., "Natural History and the Emblematic World View," in *Reappraisals of the Scientific Revolution*, ed. David C. Lindberg and Robert S. Westman (Cambridge: Cambridge University Press, 1990), 307–316.

modern reader a very different window onto the local landscapes of early modern Europe, one in many ways more open and nuanced.

NATURAL TREASURES

In 1728, the court physician Franz Ernst Brückmann published a "Subterranean Thesaurus of the Duchy of Braunschweig," or, as he subtitled it in German, "Braunschweig with its Underground Treasures and Rarities of Nature."[21] Brückmann chose to preface his book with a congratulatory poem by a colleague. This poem, composed in Latin by Albrecht Ritter, administrator of the gymnasium at Ilfeld and a fellow devotee of the mineral world, displays a striking preoccupation with the concept of natural wealth.[22] The poem begins: "Mortals often squander their treasuries [*thesauros*], for when they decide what to gather up, they find poverty in riches."[23] Here Ritter invokes a trope which, by the time he wrote, had become a commonplace in regional mineralogies: namely the perception that many lands possessed natural riches that had, up until then, been neglected or seen as worthless, but that could be redeemed if their true nature were recognized. But Ritter gave this theme a surprising twist. He discussed various different ways of seeking wealth – alchemy, mining, and trade, among others – and dismissed them all as problematic, for different reasons. In contrast to these characteristically Renaissance and Baroque modes of wealth production,[24] Ritter argued, Brückmann's project of describing natural riches was

---

[21] Franz Ernst Brückmann, *Thesaurus Subterraneus, Ducatus Brunsvigii, id est: Braunschweig mit seinen Unterirrdischen Schätzen und Seltenheiten der Natur* (Braunschweig: verlegt durch Johann Christoph Meisner, 1728), title-page. Brückmann, like many other authors of local natural history, published not only this regional mineralogy, but also a wide range of treatises on other natural products: for example, praising the virtues and offering medical explanations of the origins of various local beverages (the beer, and another drink called *Mumme*), as well as a discussion of a monster born in the vicinity. See for example his *Kurtze Beschreibung und genaue Untersuchung des Fürtrefflichen Weitzen-Biers, Duckstein genannt, Welches zu Königs-Lutter im Hertzogthum Braunschweig gebrauet* (Braunschweig: n.n., 1723); *Die Mumme scheut sich nicht/ sie will sich nicht verstecken* (Braunschweig: n.n., 1725); and *Ausführliche Beschreibung Einer seltsahmen Wunder-Geburt, Welche Eines abgedanckten Soldaten-Frau den 23. Febr. in Wolffenbüttel zur Welt gebracht* (Wolfenbüttel: Zu finden bey Johann Christoph Meißner, 1732).

[22] During the course of a career which saw his rise from "sub-corrector" to "conrector" to "prorector" of the Ilfeld *collegium*, Ritter composed numerous shorter works on nearby mineral worlds, including *Epistolica orytographia Goslariensis* (Helmstedt: litteris Buchholzianis, 1732), *Specimen I et II Oryctographiae Calenbergicae* (Sondershausen: n.n., 1741–3), and many more.

[23] Brückmann, 11.

[24] Here see Pamela H. Smith, *The Business of Alchemy: Science and Culture in the Holy Roman Empire* (Princeton, NJ: Princeton University Press, 1994). Alchemists were, in fact, called upon to serve the princely state well into the eighteenth century: see for example Janet Gleeson, *The Arcanum* (New York: Warner Books, 1998).

far more simple, sound, and likely to be productive: "You act far better and more prudently, most experienced man, when you reveal treasures to mortals which are true and beneficial, which are constant, which are true and beneficial, since they lead us to the recognition of God...."[25] Ritter ascribed the soundness of Brückmann's methods to the fact that his focus on *natural* riches led to an appreciation of God's divine Providence in having first created them.

If a single concern united the regional mineralogies of the late seventeenth and early eighteenth centuries, apart from their common focus on the mineral world, it was their preoccupation with the idea of "natural wealth." This concept recurred again and again in the ways authors described their works, in the imagery they used, and in the goals they set for themselves. Brückmann's use of the term "*thesaurus,*" or treasury, to title his book provides only one of the most obvious examples. Subtitles referred to rocks as "natural treasures" and "natural riches" (*Naturschätzen, Schätzen der Natur, Reichtum der Natur*) among other similar phrases.[26] Regional mineralogies drew repeatedly on imagery of the collection of natural abundance, alluding both to books and to territories as "storehouses" heaped full of precious objects.[27] In so doing, they drew on an early modern tradition that posited books and other cultural productions as repositories of one sort or another, declaring books to be treasuries, gardens, libraries, labyrinths, *gazophylacii*, and so forth. But they used this imagery consistently, not haphazardly, to support a central theme: that of nature as a source of previously ignored riches. In the process, the regional mineralogy acquired as its key mission the search for natural treasures even in places of seeming barrenness, proving that even apparently common natural objects could possess significant value.

As we saw above, Ritter claimed that the study of minerals might lead men to a "recognition of God." This invocation of natural theology was likewise the result of a long tradition, far too long to trace here.[28] Even in the earliest local floras, for example those of the late sixteenth century, we see the claim that the investigation of the natural creation led to a fuller knowledge of the Creator. In the very last years of the seventeenth century and the

[25] Brückmann, 11–12.

[26] See for example Georg Anton Volkmann, *Silesia subterranea, oder Schlesien mit seinen unterirdischen Schätzen* (Leipzig: Verlegts Moritz Georg Weidmann, 1720).

[27] Here see, for example, Johann Georg Liebknecht, *Hassiae Subterraneae specimen clarissima testimonia diluvii universalis* (Giessen & Frankfurt am Main: Apud Eberh. Henr. Lammers, 1730), sig.)(r: "For nature...is the richest storehouse, crammed full of such a multitude and variety of things..."

[28] Sachiko Kusukawa, *The Transformation of Natural Philosophy: The Case of Philip Melanchthon* (Cambridge: Cambridge University Press, 1995) discusses some of the origins of sixteenth-century Protestant natural theology; for a more general orientation, see John Hedley Brooke, *Science and Religion: Some Historical Perspectives* (Cambridge: Cambridge University Press, 1991).

early years of the eighteenth, John Ray's and William Derham's physico-theological works led arguments about God's wisdom and foresight in his arrangement of natural objects to a new complexity.[29] German authors wrote on the natural theology of everything from lightning to locusts, explaining the reasons why God had seen fit to place these phenomena on earth.[30] Regional mineralogies drew on similar language, though in a less developed form, presenting the diversity and singularity of specimens as continued testimony of divine Providence – and also, increasingly, of a past Deluge.[31] Natural theology thus openly encouraged talk of natural treasures, since elevating natural objects in this way simultaneously had the effect of elevating divine power.[32] Rocks could serve as testimony for many things – God's Providence, the accuracy of Biblical history, the human obligation to make use of God's gifts, the importance of a patron, the singularity of a place – all depending on the context in question.

In arguing that any natural object could be seen as valuable by nature, requiring only human attention to discover that value, and to put it to use, authors of regional mineralogies also showed strong similarities with the school of economic and administrative thought known as cameralism. Often contrasted with the mercantilist ideologies of seventeenth- and eighteenth-century Western Europe, this related economic philosophy captured the adherence of Central European policymakers during the same period. Whereas mercantilism focused on the efforts of such colonial powers as England, France, and the Netherlands to gain wealth (in particular gold bullion) through trade, cameralism instead proposed territorial self-sufficiency as a route to economic success.[33] Rather than importing expensive

[29] John Ray, *The Wisdom of God Manifested in the Works of the Creation* (London: printed for Samuel Smith, 1691); William Derham, *Physico-Theology* (London: printed for W. Innys, 1713).

[30] Sara Stebbins, *Maxima in minimis: Zum Empirie- und Autoritätsverständnis in der physikotheologischen Literatur der Frühaufklärung* (Frankfurt: Peter D. Lang, 1980).

[31] See for example Liebknecht and Giuseppe Monti, *De monumento diluviano nuper in agro Bononiensi detecto* (Bologna: apud Rossi & Socios, 1719).

[32] According to Carl Nicolaus Lange, for example, in his *Historia lapidum figuratorum Helvetiae* (Venice & Lucerne: Sumptibus Authoris, Typis Jacobi Tomasini, 1708), sig. [)bii()r, even "vulgar things" like sand could be interpreted as a sign of God's greatness.

[33] For classic overviews of mercantilism and cameralism in Europe, see Eli F. Heckscher, *Mercantilism*, translated by Mendel Shapiro (New York: Garland Publishing, 1983, c1935); D. C. Coleman, ed., *Revisions in Mercantilism* (London: Methuen, 1969); and Rondo Cameron, "Economic Nationalism and Imperialism," in *A Concise Economic History of the World* (Oxford: Oxford University Press, 1993), 130–161. On the German case in particular, see Albion W. Small, *The Cameralists: The Pioneers of German Social Policy* (Chicago, IL: University of Chicago Press, 1909); Axel Nielsen, *Die Entstehung der deutschen Kameralwissenschaft im 17. Jh.* (Jena: Gustav Fischer, 1911); Ingomar Bog, *Der Reichsmerkantilismus. Studien zur Wirtschaftspolitik des Heiligen Römischen Reiches im 17. und 18. Jh.* (Stuttgart: Fischer, 1959); Erhard Dittrich, *Die deutschen und österreichischen Kameralisten* (Darmstadt: Wissenschaftliche Buchgesellschaft, 1973); and Jutta Brückner,

foreign goods, cameralist authors argued, states would do well to further production of similar goods or alternatives at home, boosting local economies while reducing the inflow of dangerous luxuries. Cameralism, as a set of economic and political proposals, concentrated attention on the territory as a unit, and on making the fullest use of that territory's own resources, natural, human, and otherwise.

Developed in various courts of the Empire by such thinkers and entrepreneurs as Johann Joachim Becher, cameralist ideas were taken up by courts throughout the Holy Roman Empire, as cash-strapped rulers sought to make full use of their territories' resources to provide badly-needed funds for their treasuries. Books such as Johann Rudolf Glauber's *Teutschlands Wolfahrt* (1656), Veit Ludwig von Seckendorff's *Teutscher Fürsten-Staat* (1665), and Becher's *Politischer Discurs von den eigentlichen Ursachen deß Auf- und Abnehmens der Stadt, Länder und Republicken* (1673) promoted the intensive development of local enterprise to make use of raw materials at hand.[34] Examples of proposed projects included the construction of factories and manufactures to take advantage of concentrations of particular resources in territories; the establishment of mines for the extraction of particular resources; and experiments with ways of increasing agricultural productivity.[35] Though some of the earliest versions of these projects were no more than utopian sketches, by the late seventeenth century they were increasingly proving valuable tools for court treasurers and administrators faced with the task of maximizing revenue, to offset increasing expenditures for armies and for the upkeep of courts in an age of emulation and luxury. A key consideration thus became the maintenance of self-sufficiency in economic, and natural, matters. In keeping with this, princes began to pay closer attention to the contents of their territories.

In the regional mineralogies of the late seventeenth and early eighteenth centuries, striking affinities with the themes of the early cameralist literature may be seen. Just as cameralist treatises argued for the necessity of economic

Staatswissenschaften, Kameralismus und Naturrecht. Ein Beitrag zur Geschichte der politischen Wissenschaft im Deutschland des spaten 17. und frühen 18. Jahrhunderts (Munich: Beck, 1977).

[34] Johann Joachim Becher, *Politische Discurs, von den eigentlichen Ursachen, des Auff- und Abnehmens der Städt, Länder und Republicken* (Frankfurt: J. D. Zunner, 1673); Johann Rudolf Glauber, *Teutschlands Wolfahrt* (Amsterdam: gedruckt bey Johan Jansson, 1656); Veit Ludwig von Seckendorff, *Teutscher Fürsten Stat* (Frankfurt: in Verlegung Thomas Mattiae Gotzens, 1665).

[35] See Ulrich Troitzsch, *Ansätze technologischen Denkens bei den Kameralisten des 17. und 18. Jahrhunderts* (Berlin: Duncker & Humblot, 1966); Pamela H. Smith, *The Business of Alchemy*; and R. Andre Wakefield, "The Apostles of Good Police: Science, Cameralism, and the Culture of Administration in Central Europe, 1656–1800," Ph.D. dissertation, University of Chicago, 1999. On the German courts more generally, see Adrien Fauchier-Magnan, *The Small German Courts in the Eighteenth Century*, translated by Mervyn Savill (London: Methuen, 1958).

self-sufficiency, for example, regional mineralogies took exhaustive inventory of territories' mineral contents, arguing that in each case all the necessary mineral kinds were represented, so as to enable freedom from dependence on other areas' natural products.[36] Likewise, regional mineralogies echoed cameralist insistence on the necessity of transforming natural objects, in one way or another, to release their value. Hidden in the earth, a metal, a fossil, or other mineral was useless; only when made public could the "public good" be served.[37] As has recently been shown, the natural-historical projects of the Swedish professor Linnaeus were shaped in large part by his involvement in Swedish cameralist projects during the mid-eighteenth century.[38] The case for such direct influence in the Holy Roman Empire of the mid-seventeenth century onwards is more indirect. At this stage we see rather an affinity of ideas, a resonance of themes.[39] But the similarities may not be dismissed. Perhaps the best way to proceed is simply to examine the role that concepts of "natural wealth" played, with all their richness of connotation and reference, in shaping the direction of the regional mineralogy.

Take, for example, the issue of hierarchy. One of the features shared by both mineralogical and cameralist literature was their common interest in the revealing of objects previously seen as useless, yet now seen as possessing value. Just as certain hierarchies had traditionally been popularly imputed to the plant world – with some plants seen as nobler and more refined, while others were viewed as coarser and lower[40] – even more definite hierarchies existed in the world of minerals and metals. Nowhere may this be seen more clearly than in the mining literature, where "noble metals" such as gold and silver warranted reverential attention, with "common" or "base" substances

[36] See for example the physician Johann Heinrich Schütte's insistence, in his *Oryktographia Jenensis, sive fossilium et mineralium in agro Jenensi brevissima Descriptio* (Leipzig: Sumptibus Josephi Wolschendorfii, Typis Hermannianis, 1720), sig. (*)v, that his work would be of not only medical but "economic" purposes.

[37] It should be pointed out, though, that not all regional mineralogies actually ended up being published, though presumably most were; the Handschriftensammlung of the Universitätsbibliothek Erlangen contains a manuscript copy of a work by Johann Wilhelm Kretschmann, a physician in the Bavarian town of Hof near Bayreuth, entitled *Marchionatus Brandenburgico Baruthini Subterranei Specimen primum oder ersten Versuch einer Beschreibung des Unterirdischen Marggraffthums Brandenbürg-Bayreuth*, apparently composed in 1736 and never published (Ms. B248). A later version of this manuscript from 1741 finally ended up being published in facsimile, after an interval of several centuries: see Johann Wilhelm Kretschmann, *Sammlung zu einer Berg-Historia des Marggrafthums Brandenburg-Bayreuth*, ed. Dieter Arzberger (Selb-Oberweissenbach: Arzberger, 1994).

[38] Lisbet Koerner, *Linnaeus: Nature and Nation* (Cambridge, MA: Harvard University Press, 1999).

[39] See the next chapter, however, for a case which demonstrates a more direct connection between interests in cameralist theory and natural history: namely that of Urban Gottfried Bucher, the author of a natural history of Saxony who also wrote a favorable synopsis of Johann Joachim Becher's works.

[40] Allen J. Grieco, "The Social Politics of Pre-Linnean Botanical Classification," *I Tatti Studies: Essays in the Renaissance* 4 (1992): 131–149.

dismissed in a sentence, or not mentioned at all. Regional mineralogies often adopted this terminology of "noble" and "base." Yet even when they used this wording, they often did so in ironic fashion, insisting, for example, that "common" minerals were every bit as much worthy of interest as "noble" ones, or that noble ones were not, in fact, so noble after all.

Regional mineralogies thus, by treating a wide range of common and uncommon substances equally, may be seen as tending towards a leveling or equalizing effect. Like local floras, they set all natural objects on the same plane, collapsing hierarchies so as to be able to examine and extol the merits of each and every natural production. This collapsing of hierarchies has often been seen, in the realm of political history for example, as a feature of Enlightenment thought. Yet we see a similar phenomenon occurring in the description of the *natural* world much earlier. Early modern natural historians and natural philosophers alike had long since fashioned a commonplace out of Aristotle's famous contention, in his *De partibus animalium*, that it was as important to study the "meaner" creatures as those that were more prestigious, since "in all natural things there is somewhat of the marvellous."[41] The sixteenth-century herbalist Hieronymus Bock, for example, had chosen to begin his *New Kreutterbuch* with a listing of the humble thistle, rather than of any rarer or more elevated plant.[42] Likewise, among the regional mineralogists, we see Johann Jakob Baier cheerfully conceding that there were no "noble" metals in the Nuremberg area.[43] In the imperial free city of Nuremberg, free of any resident court, Baier could make such a claim, while writers explicitly aiming to please princes operated under different constraints. Nonetheless Peter Wolfart, writing for the *Landgraf* of Hessen, ended up dividing the rocks he discussed into "precious" and "vulgar, or less precious," suggesting something less than an absolute hierarchy. Even "vulgar" rocks might themselves possess something "precious" about them, and Wolfart set out to prove this through his ample description of their value and uses.[44]

In this evolving constellation of ideas, where every different rock was seen as possessing an intrinsic value of its own, new emphasis was placed on sheer *variety*. Certainly an interest in the "diversity" of the natural world

[41] Aristotle, *Parts of Animals*, translated by E. S. Forster (Cambridge, MA: Harvard University Press, 1955), 98–101.
[42] Hieronymus Bock, *New Kreütterbuch* (Strasbourg: durch Wendel Rihel, 1539), sig. [4]v.
[43] Baier, 94. Baier chose to list metals at the end of the book, rather than giving them pride of place as in the mining literature. Even more strikingly, Jakob von Melle, also a citizen of an imperial free city, admitted in his *De lapidibus figuratis agri litorisque Lubecensis* (Lübeck: typis Struckianis, 1720), 43, that there were numerous kinds of specimens he could *not* find in the neighboring region, and went so far as to list them, possibly in the hopes that his correspondents from areas where these were more abundant might send him specimens (see the final section of this chapter).
[44] Peter Wolfart, *Historiae naturalis Hassiae inferioris, Pars prima...i.e. Der Natur-Geschichte der Nieder-Furstenthums Hessen Erster Theil* (Kassel: gedruckt bey Heinrich Harmes, 1719), 17 and 22.

was long-standing in European culture, dating back at least to the medieval period; but much of this interest seems to have been linked to the concept of the "diverse" as strange, as a quality to be found in the East, rather than at home.[45] More recently, local florists had concerned themselves with such issues as well, on occasion counting the number of different species they were able to list, and boasting about their sheer quantity, just as they had boasted about the number of exotic species they had successfully established in their botanical gardens. But this concern with the "diversity" of nature's productions became especially marked in the regional mineralogy. Authors here boasted not of sheer numbers of natural kinds, but rather of the individuality of each specimen they found. Whereas plant species ensured that each plant found would be reasonably close to type, the individual character of each "figured stone," requiring illustration to demonstrate this individuality, resulted in the celebration of sheer "difference" for its own sake. In this view, even the mere existence of a sufficient variety of specimens, however "humble" or seemingly valueless, could help establish the region in which they were found as possessing significant natural wealth, one independent of human monetary wealth.

And just as the diversity of previously ignored natural objects came to acquire new importance in the regional mineralogy, so too did that of places and entire environments. Writers of regional mineralogies had long since headed up the slopes of hills and mountains, in the footsteps of writers of local floras, to discover not only the wild and varied flora that often grew there, but also the various inorganic materials to be found on or beneath the surface. Moritz Hoffman's fascination with the Moritzberg, the highest point near Altdorf, was echoed by other authors, for whom mountains and other geological outcrops constituted far more rewarding terrain than the level ground on which cities and towns had often been built. A particularly vivid example of the exploration of new terrains may be seen in the case of caves. Whereas miners had long worked amidst the terrors and mysteries of underground passageways, developing their own protective folklore in the process, naturally-formed caves were in the seventeenth century still often shunned. Gateways to subterranean worlds, and possibly hells, they were also dangerous places to visit. Those that had been canvassed in search of metals, and subsequently abandoned, commonly were the sites of legends about demons and evil spirits.[46] Yet over the course of the seventeenth and

[45] See Katharine Park, "The Meanings of Natural Diversity: Marco Polo on the 'Division' of the World," in Edith Sylla and Michael R. McVaugh, eds., *Texts and Contexts in Medieval Science: Studies on the Occasion of John E. Murdoch's Seventieth Birthday* (Leiden: Brill, 1997), 134–147. I am also indebted to Londa Schiebinger's discussion, in *Nature's Body: Gender in the Making of Modern Science* (Boston, MA: Beacon Press, 1993), of natural history as a science of *difference*.

[46] See Baier, 10 for an evocation of the horrors of caves, with their "perpetual night" and "very dense shadows."

eighteenth centuries we see a gradual introduction of a tourism of sorts in these precarious locations. By the early eighteenth century, academicians such as Albrecht Ritter took their students on spelunking expeditions.[47] And the learned felt it a duty to check out local caves for any possible treasures – of the learned kind. For there was no question of a search for gold or silver here. Professors searched caves as much for the marvelous forms of their stalactites and stalagmites, as for any explicitly "useful" minerals. All were presented as equally of interest.

Johann Jakob Baier's *Oryktographia Norica* (1708) illustrates the ways in which the language of natural riches developed in the regional mineralogy could serve broader notions of utility. The region around Nuremberg, Baier maintained, was indeed naturally fertile, though more for agriculture than for mining, owing to its sandy soils.[48] But this fertility needed to be recognized and harnessed, in order for its potential advantages to be exploited. Baier went through all possible "fossil" categories one by one, demonstrating that they had either been located in the Nuremberg area or "were yet to be found"; in other words, he held out hope that the Nuremberg area would be found to possess a full complement of every kind of natural object, if only the search were conducted vigorously enough.[49] He displayed a keen interest in the "uses" of natural objects, recounting folk and popular uses as well as those in mining, industry, and medicine, on occasion suggesting new uses of his own.[50] In calling attention to these "uses," however, Baier seems not to have had merely economic utility in mind, but rather a broader concern for recognizing places' and objects' hidden potential. For Baier, as for many other authors of regional mineralogies, the endeavor to locate and publicize previously unknown natural riches was inseparable from the larger project of attempting to develop and fulfill their natural potential.

EXCAVATING WÜRZBURG

To examine these issues in greater depth, I would like to retell and reframe an example of an early eighteenth-century regional mineralogy already familiar to many historians of science: the history of a volume titled the *Lithographiae Wirceburgensis* (1726), or, literally, "rock descriptions of Würzburg."[51]

---

[47] For more on caving, see also Baier, 23 and Lachmund, 62.   [48] Baier, 1–2.

[49] Baier, 49, 93.   [50] Baier, 15, 24, 92, 94.

[51] Johann Bartholomäus Adam Beringer, *Lithographiae Wirceburgensis, ducentis lapidum figuratorum, a potiori insectiformium, prodigiosis imaginibus exornatae specimen primum* (Würzburg: apud Philippum Wilhelmum Fuggart, 1726). This volume has been reprinted and translated as *The Lying Stones of Dr. Johann Bartholomew Adam Beringer, being his Lithographiae Wirceburgensis*, translated and annotated by Melvin E. Jahn and Daniel J. Woolf (Berkeley: University of California Press, 1963). I will use their translation of this work unless specified otherwise. In addition, I will follow their sequential narration of events on 1–7 and 125–130.

What has won this work its place in history is the fact that the figured stones it described turned out to be fakes. To historians of science, this work has thus been of interest primarily as a morality tale, or, more recently, as a case study in scientific fraud.[52] But if the *Lithographiae Wirceburgensis* is examined as belonging to the genre of the regional mineralogy – and this is indeed the way in which its author chose to frame it – a very different set of concerns rises to the surface.

On May 31, 1725 Johann Bartholomäus Adam Beringer, a court physician and professor of philosophy and medicine at the university of Würzburg, received a delivery of unusual fossils from several local youths he had hired to hunt for fossils. Purportedly dug up on a nearby hill, Mt. Eivelstadt, where Beringer had often before found interesting fossils, these fossils were *truly* interesting, displaying the figures of worms and the sun's rays, seemingly captured almost entire in the rock.[53] Delighted with these finds, Beringer combed the mountain himself in search of further such fossils, and urged his hired youth to search for more. Over the course of the next several months he amassed a considerable collection of these extraordinary specimens (for some examples, see Figure 8). Eager to make known his finds, he decided to publish them in local natural history form, that is to say, to frame them as a "Würzburg Lithography," or regional mineralogy, in which he presented these natural objects as emblematic of and reflecting on the nature of Würzburg itself.[54]

However, his task had already become problematic. Rumors had already begun to circulate that the stones' striking images were the work not of nature, but of artifice: in particular, the artifice of some modern forger, who had planted these stones as a hoax to deceive Beringer. Indeed, the images on some of the stones seemed to confirm this impression: for example, the apparently Hebrew letters appearing on several specimens (see Figure 9). Beringer at first resisted the rumors, arguing that if this were a hoax, someone would have had to have gone to an extraordinary amount of trouble to bring it about. Ultimately, he was forced to the conclusion that the rocks *were* fraudulent – but by this point it was too late, for the book had already been published. He is said to have frantically attempted to have all copies of the book burned – but it was too late for this as well, since he had already sent out numerous copies to scholars all over Europe.

---

[52] Jahn and Woolf, eds., 126–127, outline the development of this morality tale during the eighteenth century. Other accounts, focusing specifically on the way Beringer's case *has* been used as a morality tale, include Hans Franke, *Die Würzburger Lügensteine. Tatsachen, Meinungen und Lügengespinste uber eine der berühmtesten geologischen Spottfälschungen des 18. Jahrhunderts* (Würzburg: Schöningh, 1991) and Stephen Jay Gould, "The Lying Stones of Marrakech" in his book of the same title (New York: Harmony Books, 2000), 9–26.

[53] Jahn and Woolf, eds., 2 and 129.   [54] Jahn and Woolf, eds., 3.

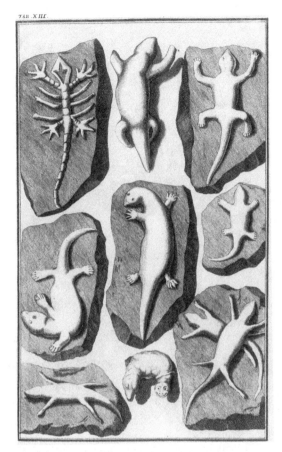

Figure 8. Figured Stones. Though these specimens may seem very crude to us today, many seventeenth- and eighteenth-century mineralogical works contained similarly striking images of the apparently lifelike forms to be found in many rocks. From Johann Bartholomäus Adam Beringer, *Lithographiae Wirceburgensis* (Würzburg: apud Philippum Wilhelmum Fuggart, 1726). From the collections of the Ernst Mayr Library, Museum of Comparative Zoology, Harvard University.

So who was ultimately responsible for the debacle? The story that ended up being told by eighteenth-century naturalists, and that was subsequently picked up by historians of science, was a morality tale: one in which a researcher, deluded by his own ambitions and by a fundamental misconception of the nature of reality, had let himself be deceived by the pranks of his very own students, who had planted these rocks to make fun of their professor.[55] But the story is more complicated than that, as shown by

[55] Jahn and Woolf, eds., 126–127; see also p. 3 for more on "popular history's" version of the story.

Figure 9. Figured Stones with Hebrew Letters. Though not all of the symbols here
appear to be from the Hebrew alphabet, enough are – including the Tetragrammaton, the sacred Hebrew name of God that appears at the top center – to make
the phenomenon seem like more than mere chance. Note the odd variations on the
Tetragrammaton, though, top left and center. From Johann Bartholomäus Adam
Beringer, *Lithographiae Wirceburgensis* (Würzburg: apud Philippum Wilhelmum
Fuggart, 1726). From the collections of the Ernst Mayr Library, Museum of Comparative Zoology, Harvard University.

the rediscovery earlier in this century of a set of court proceedings initiated
by Beringer after the fiasco.[56] Here it came out that Beringer's tormentors
were no mere group of mischievous students, but were rather professional
colleagues of his at university and court, whose rivalry had been inflamed

[56] See Heinrich Kirchner, "Die Würzburger Lügensteine im Lichte neuer archivalischer Funde,"
Zeitschrift der Deutschen Geologischen Gesellschaft 87, 9 (Nov. 1935): 607–615. English
translations of the judicial proceedings may be found in Jahn and Woolf, eds., 130–141.

by his perceived successes in natural history.[57] One of these men left town in disgrace, and the other one died shortly thereafter, thus imposing at least some closure on this sordid episode in the history of scientific error.[58]

It is useful, though, to look at Beringer's book as he himself chose to frame it: namely, in the style of the regional mineralogy. In fact, the *Lithographiae Wirceburgensis* shares several key features with other early eighteenth-century regional mineralogies. First, despite Beringer's status as a university professor, the work in fact illustrates the spread of local natural history beyond the confines of the university. For Beringer was not merely a professor; he was also the chief personal physician (*Leibarzt*) of the Prince–Bishop of Würzburg, and one of his chief advisors (*Hofrat*) as well. As such, he joined the ranks of an increasing number of court officials and bureaucrats who saw in these works a new way to convince their princes and patrons of the usefulness of their services.[59] Regional mineralogies often moved beyond limits earlier provided by the garden and the town, going well beyond the "local" in their endeavor to survey the contents of entire territories, and often suggesting ways in which the natural productions of these territories could be put to the greater use and glory of their territorial superiors.[60]

These tendencies can be seen in the *Lithographiae Wirceburgensis* as well. Rather than dedicating his book to the traditional mix of university officers and other notables, Beringer had the book dedicated instead to his lord and master, the Prince–Bishop of Würzburg.[61] Phrased in the language of court culture, the dedication attempted to convince the prince of the very real significance of the natural world for his rule. The natural treasures he had found on the mountainside, Beringer argued, revealed that Würzburg was blessed with substantial natural wealth, if one only knew how to find it. Though the prince's power ultimately came from God, God had chosen to reveal this through the wonderful workings of nature, thus demonstrating that nature herself served as a basis and a foundation for the prince's rule. The book's frontispiece provides a telling example (see Figure 10). Here, we see

[57] Jahn and Woolf, eds., 3.   [58] Jahn and Woolf, eds., 139.

[59] See Vivian Nutton, ed., *Medicine at the Courts of Europe, 1500–1837* (London: Routledge, 1990) and Andrew W. Russell, ed., *The Town and State Physician in Europe from the Middle Ages to the Enlightenment* (Wolfenbüttel: Herzog August Bibliothek, 1981).

[60] A growing body of literature discusses the patronage of early modern science in the court context; see for example Mario Biagioli, *Galileo, Courtier: The Practice of Science in the Culture of Absolutism* (Chicago, IL: University of Chicago Press, 1993) and Bruce T. Moran, ed., *Patronage and Institutions: Science, Technology and Medicine at the European Court, 1500–1750* (Woodbridge, UK: Boydell, 1991).

[61] Beringer, dedication; Jahn and Woolf, eds., 11–14. Although the book appears in the format of an academic dissertation, with Beringer as *praeses* and a student of Beringer's by the name of Georg Ludwig Hueber (a shortened form of Hueberg) as respondent, it appears that, as was common in these cases, Beringer composed the work himself (see Jahn and Woolf, eds., 6–7).

Figure 10. Frontispiece to Beringer's *Lithographiae Wirceburgensis*. Atop the heap of fossils appear the arms of the Prince–Bishop of Würzburg. The city of Würzburg itself lies in the background, center left. From Johann Bartholomäus Adam Beringer, *Lithographiae Wirceburgensis* (Würzburg: apud Philippum Wilhelmum Fuggart, 1726). From the collections of the Ernst Mayr Library, Museum of Comparative Zoology, Harvard University.

the mountain where Beringer conducted his excavations, heaped high with fossils showing images from across the entire spectrum of natural diversity: a crayfish, a shell, a lizard, a shooting star, etc. Resting on this hill, and on what we would thus now see as a *very* fragile foundation, were the symbols of the prince's power itself. Through the fossils, nature had created for the prince "a new monumental pyramid to your glory – a pyramid impervious to the scorching sun of the summer and the blasts of the winter's storms, because fashioned from flint and marble."[62] In short, Beringer and others came to present the investigation of local nature as extremely relevant, both practically and symbolically, to those responsible for the inner functioning of a territory or state.

Throughout the *Lithographiae Wircebergensis*, Beringer's interest in the concept of "natural wealth" reflects that present in the genre of the regional mineralogy as a whole. Like other authors, he used his treatise to argue for his region's fertility not only in agricultural produce, or in numbers of inhabitants, but also in natural productions more generally, ones that may have previously been ignored, but that were now for the first time seen as possessing a value. Areas previously seen as barren and sterile were now regarded as harboring previously unseen natural riches. Beringer's mountain was one such place. According to him, the fertility of the rest of Franconia was fairly well established. "...the blessed crops fill the granaries, not only of our country, but of other regions as well. Rich pastures nourish vast flocks of cattle. Streams and rivers, descending through the valleys with placid current, almost match their waves with fishes.... Salt sources gush forth. Healthful and medicinal pools boil. In other parts of our meadows and woods there flourishes a rich and rare harvest of botanical treasures."[63] In this glowing portrait, just about the only aspect of nature that is missing is the mineral world.[64] But Beringer set out to show that what was apparently missing could, in fact, have been there all along. He described the mountain thusly: "...it bristles with fragments of rock protruding from it in wild disorder. It has no clusters of trees, only mere shrubs. It is stark, bald, and sterile."[65] Yet through his excavations Beringer prided himself on having redeemed this apparently barren mountain as a veritable "horn of plenty" (*cornucopia*), thereby revealing the territory of Würzburg as now whole and complete in its natural aspects, possessing everything it could possibly require.[66] Other writers of regional mineralogies were to make similar claims in abundance about their own territories, systematically going through lists of important natural substances one by one and attempting to argue that their own territory possessed quite satisfactory quantities of each of these. Through their focus on

[62] Beringer, dedication; Jahn and Woolf, eds., 13.
[63] Beringer, 4; Jahn and Woolf, eds., 19–20.   [64] Beringer, 4; Jahn and Woolf, eds., 20.
[65] Beringer, 19–20; Jahn and Woolf, eds., 34.   [66] Beringer, 5; Jahn and Woolf, eds., 20.

completeness and self-sufficiency, these regional mineralogies sought to pro-
claim the territory itself as a "natural" unit, arguing that nature herself had
provided, conveniently within the boundaries of a given territory, everything
needed to sustain its self-sufficient existence.

The degree to which Beringer phrased his book in this language of "natural
riches," and the degree to which others perceived the project in this way, may
be seen in an anonymous parody written and circulated in manuscript at the
time. The parody began by playing on Beringer's obsession with the mineral
world. "Take heed! There's something stony in the air. For if all others are
silent, the very stones shall cry out, manifesting the obscure and obscuring
the manifest.... From the very heavens he has stoned us, this peerless Doctor
J. B. A. B."[67] From here, the parodist moved on to mock Beringer as having
made exaggerated claims for the importance and relevance of natural history
to the princely court: "The wonders of Nature he displays, and throws open
his stone treasury, an unexampled storehouse of curios, notable for its bibicu-
lous lavishness. Hither and yon the tidings fly. Royal treasurers are enticed
to repair in all haste and enrich themselves – for a price. Bronze etchings,
pressed to paper, do duty where the stones cannot be heard, and lo! we have
paper stones! and these aspire to have a stony hardness, sadly lacking in the
originals."[68] In particular, the parodist suggested that Beringer's claims for
the usefulness of mineral knowledge seemed to compete with those of mining and alchemy. "Some... see in him a millstone, still others, a cornerstone.
Some would see him raining pearls; others, emeralds. And those there are
who look to him as an alchemist, for well they know the great power that is
in stones, in words, and in herbs. That of stones is a sacred power, and great
is the power they yield. For who will gainsay that stones, consorting with the
stars, partake of the very power of these?"[69] The parodist, triumphantly (if
somewhat misleadingly) noting that "the quarry-man deceived the doctor,"
concluded that "The stone mines are at last exhausted."[70]

This leads to a final point. Although, as we have seen, Beringer in many
ways constructed his work very much for a "local" audience, namely that
of the Würzburg court, nonetheless he simultaneously aimed his work at
an international audience of scholars. Beringer discussed the Würzburg fossils not solely on their own terms, but rather compared them with similar
specimens found throughout Europe. To do so, he turned to regional mineralogies written not only by German authors, but by a wide range of other
European naturalists, whose works he cited extensively. The famous English

[67] For a brief discussion and translation of this parody, see Melvin E. Jahn, "A Further Note
on Dr. Johann Bartholomew Adam Beringer," *Journal of the Society for the Bibliography of
Natural History* 4, 3 (1963): 160–161; I use Jahn's translation here.
[68] Jahn, 160.   [69] Jahn, 160.
[70] Jahn, 160. Actually, of course, it was Beringer's peers who deceived him, though how great
a role the teenage stone-cutters in fact played we shall probably never know.

mineralogist Woodward was a frequent presence in his text, as were many others.[71] Beringer drew on the growing literature of local natural history because it presented an opportunity for *comparison*, and thus for further definition of his own local specimens. Beringer referred repeatedly to the "gratifying exchange" (*jucundo commercio*) that correspondence had afforded him, urging others to compare their local fossils with those elsewhere.[72] At the end of the *Lithographiae Wirceburgensis*, he urged scholars elsewhere to compare his findings with theirs.[73] By turning to the works of others, who were in the process of establishing themselves as specialists in the local natural history of their own territories, Beringer was likewise able to establish his own role as purveyor and interpreter of Würzburg's natural wealth.

Beringer may thus be seen as belonging to an international network of specialists on local nature, which had gradually developed since the Renaissance. This international network did not confine itself either to rocks or to plants, but rather mixed classes and categories of specimens to create new categories of expertise. Beringer was a full participant in this international network. As his correspondence reveals, and as the will show, he and others made use of their geographical situation to establish themselves as specialists in a number of natural domains. Not only did Beringer attempt to set himself up as a knowledgeable provider of Würzburg's mineral wealth, he also used his situation as director of the Würzburg botanical garden to orchestrate a series of trades and exchanges, in which he acquired and supplied seeds and rocks from throughout the Germanies, Europe, and beyond. In so doing, he participated in a form of exchange characteristic of this period: exchange not for money, but rather in kind, i.e. exchange of one natural object for another. As naturalists traded specimens, they acquired a better knowledge of the kinds of natural wealth distinctive to their own territories – and they used this natural wealth as a form of coin for acquiring knowledge.

DEALING IN THE LOCAL

As the last set of examples will have made clear, the production of regional mineralogies, as in the practice of local natural history more generally, was far from an entirely solipsistic endeavor – that is to say, it involved the making of connections and comparisons far beyond the "local." Just as Beringer compared his "fossils" with those discussed by numerous other European mineralogical authors, so too did compilers of regional mineralogies throughout the Empire examine their own local specimens in the

---

[71] In his very first chapter, for example, Beringer took the time to compose "a brief alphabetical synopsis of Lithographers" in list form (Beringer, 14–17; Jahn and Woolf, eds., 29–31).

[72] Beringer, 4 and 87–88; Jahn and Woolf, eds., 20 and 101–102.

[73] Beringer, 95; Jahn and Woolf, eds., 109.

light of those from other lands. Presenting themselves as specialists in the rocks of their own territories, they traded these – and the books they wrote about them – for specimens from elsewhere. In this way, they took part in an exchange economy, not so much the economy of monetary exchange that had for centuries been gaining strength in Europe, but rather an economy of exchange-in-kind, or barter, in which diverse natural objects came to be seen as substances of value in their own right.[74] By exploring the correspondence networks that grew up around the exchange of local specimens, and of localist works in general, this section will show how naturalists used the local "natural riches" of their own and neighboring regions to reach beyond purely local boundaries, and to situate themselves in a set of larger contexts.[75]

Fortunately for the historian, the Trew collection of manuscripts at the Universitätsbibliothek Erlangen has preserved much of the correspondence that made this kind of exchange possible. Christoph Jakob Trew was a Nuremberg physician of the mid-eighteenth century, with a passion for botany and for collecting letters and documents related to the history of medicine. His collection eventually encompassed correspondence ranging from the late sixteenth century through his own day.[76] Though some of the scholars whose papers he acquired were well-known (for example, Trew made a point of buying up the papers of the renowned Camerarius family), most of the physicians represented in his collection were relatively obscure. As likely to be court physicians or town physicians as university professors, many of these confined their publishing activities to comments upon local medical and natural phenomena, whether by writing local floras or mineralogies, or by writing up descriptions of extraordinary births or other unusual events or characteristics of their area. Of the authors I have discussed so far, many left traces in Trew's collection.

Franz Ernst Brückmann, for example, whom we have already encountered earlier in this chapter, corresponded directly with Trew. Brückmann may be seen as someone who built much of his career on the exchange of

74 For a discussion of early modern systems of exchange in the Italian context, see Paula Findlen, "The Economy of Scientific Exchange in Early Modern Italy," in *Patronage and Institutions: Science, Technology, and Medicine at the European Court, 1500–1750*, ed. Bruce T. Moran (Woodbridge, UK: Boydell, 1991), 5–24.

75 I also discuss this phenomenon in my article "From the Alps to Egypt (and Back Again): Dolomieu, Scientific Voyaging, and the Construction of the Field in Eighteenth-Century Europe," in *Making Space: Territorial Themes in the History of Science*, ed. Crosbie Smith and Jon Agar (London: Macmillan, 1998), 39–63.

76 The collection is catalogued in E. Schmidt-Herrling, *Katalog der Handschriften der Universitätsbibliothek Erlangen, V. Die Briefsammlung des Nürnberger Arztes Christoph Jacob Trew (1695–1769)* (Erlangen: Universitätsbibliothek, 1940). For an interpretation of the role of correspondence in Trew's own career, see Thomas Schnalke, *Medizin im Brief. Der städtische Arzt des 18. Jahrhunderts im Spiegel seiner Korrespondenz* (Stuttgart: Steiner, 1997).

letters, many of which he subsequently printed in his *Epistola itineraria*.[77] In his letters to Trew, he presented himself as an expert supplier of minerals and "figured stones" from Braunschweig and neighboring areas, eventually encompassing much of northern Germany. He engaged Trew in a set of exchanges, inquiring about whether Trew could send him fossils from Altdorf and nearby Eichstädt in return for some of Brückmann's northern specimens. In the process, he also discussed the publication of his own mineralogical works, and tried to publicize them to Nuremberg *Curiosi*. In this multiple set of transactions, involving the frequent exchange of small parcels ("packets") containing *naturalia*, books, and occasionally money, Brückmann established himself as an expert on the rocks of his area, and parlayed this expertise into a means of acquiring specimens from other lands.

In his very first letter to Trew preserved in the collection, Brückmann announced that he would be sending him "several petrifications of our land" via a mutual friend.[78] He inquired whether Trew could possibly send him some specimens from the region around Nuremberg in return: "If you ever acquire any attractive fishes or a crab or sea-star from Eichstädt, or the agreed-on land-snails in rock from the Württemberg area, please send them via Ebermayer [the mutual friend] or someone else."[79] Subsequent letters show Brückmann developing his relationship with Trew further, sending him more natural objects from his parts,[80] and making further requests for fishes from Eichstädt and fossils from Altdorf ("I'd really like another portion of those Altdorf figured stones...").[81]

Nor was his interest limited to the exchange of *naturalia*. Brückmann's requests for fossils were soon supplemented by discussions of his publishing endeavors, and appeals for Trew and his acquaintances to subscribe to a work he had written entitled *Beschreibung aller Bergwerke* (Description of all Mines). Here he went into numerous financial details about the subscription scheme.[82] Two letters down the line, he expressed his wonder that no one from Nuremberg had yet ordered the book, and asked Trew for help in publicizing it.[83] Not long after, he confided that the book was indeed not selling well, and asked if Trew could do him the favor of inserting an "*Avertissement*" for it in his *Commercium Literarium*, a Nuremberg-based journal Trew edited. In the same letter, he took care to inform Trew of the progress of his next publishing venture: "Right at the moment I'm having *Subterranean Braunschweig* printed, and there are already two gatherings

<hr>

[77] These were printed in numerous editions; for an even hundred of them, see Franz Ernst Brückmann, *Centuria epistolarum itinerarium* (Wolfenbüttel: n.n., 1742).

[78] Brückmann mss., Universitätsbibliothek Erlangen, letter of 30 May 1725, 2. I will hereafter cite this correspondence as "Brückmann mss."

[79] Brückmann mss., letter of 30 May 1725, 2.   [80] Brückmann mss., letter of 19 Dec. 1727.

[81] Brückmann mss., letters of 23 May 1728 and 2 Aug. 1729.

[82] Brückmann mss., letter of 12 Feb. 1727.   [83] Brückmann mss., letter of 26 May 1727.

ready...."[84] Clearly, Brückmann hoped his mineralogical exploration of Braunschweig might appeal to a similar group of people as those he envisioned as the audience for his mining book: "lovers of mines and minerals, possessors of cabinets of art and *naturalia*, and book-collectors in Nuremberg."[85] Through the remainder of Brückmann's correspondence with Trew, during the course of which he sent him a picture of a hermaphroditic carp from Goslar and ordered numerous glass eyes from Nuremberg,[86] he repeatedly mentioned his Braunschweig *Thesaurus Subterraneus* amidst his discussions of local *naturalia* and books. In one case, he responded to an inquiry of Trew's about an egg-stone by broadly hinting that "the piece itself is not to be had any longer," but that Trew could find a description of it in the Braunschweig *Thesaurus subterraneus*.[87] Brückmann's correspondence with Trew thus reveals an exchange economy on many levels: of natural objects, of books and images describing them, and (only occasionally) of actual money. None of these levels was primary; rather, all combined together to form a dense web of interconnections in which local objects, by being identified as local, were given a value and subsequently sent out to the outside world.

An examination of the correspondence of Johann Bartholomaeus Adam Beringer (the same who proved victim to the lying stones of Würzburg) shows yet another way in which the exchange of natural objects could develop. Here Beringer turned to Trew and to Johann Georg Volckamer (another Nuremberg botanist), not as providers of local minerals, but as sources of exotic seeds from Amsterdam.[88] For Beringer, prior to his unfortunate lithographic activities, had in 1704 recently been appointed director of the newly-founded botanical garden in Würzburg. He thus had an urgent need for plants to put in it, and a desperate lack of any seeds he could use to exchange for new ones. Beringer was acutely aware of "my poverty and lack of provisions" in this matter.[89] Though he proudly told Volckamer that he had already begun to develop his own connections in Amsterdam – "there

---

[84] Brückmann mss., letter of 24 June 1727.   [85] Brückmann mss., letter of 19 Dec. 1727.
[86] Brückmann mss., letter of 12 Oct. 1733.   [87] Brückmann mss., letter of 11 April 1734.
[88] Johann Georg Volckamer (sometimes labelled Johann Georg Volckamer II, so as to avoid confusion with his father of the same name) was a well known botanist-physician and the author of a *Flora Noribergensis, sive Catalogus plantarum in agro Noribergensi tam sponte nascentium, quam exoticarum* (Nuremberg: sumtibus Michaellianis, 1700), which, as its title suggests, included plentiful references to exotic plants as well as local ones. His brother, merchant Johann Christoph Volckamer, shortly thereafter drew on his extensive connections abroad to compile and publish a sumptuous *Nürnbergische Hespebibes, oder, Grundliche Beschreibung der edlen Citronat- Citronen- und Pomeranzen-Fruchte, wie solche in selbiger und benachbarten Gegend, recht mogen eingesetzt, gewartet, erhalten und fortgebracht werden* (Nuremberg: bei dem Authore, 1708–1714), instructing Germans on the cultivation of citrus trees for their gardens and *orangeries*. Beringer's correspondence with Johann Georg thus enabled him to draw on the resources and international networks of his brother as well.
[89] Beringer mss., Universitätsbibliothek Erlangen, letter of May 10, 1704. I will from now on cite this correspondence as "Beringer mss."

are some seeds, curious overseas ones with their foreign names, which I've carefully arranged to be sent to me from Holland . . . namely from the Medical Garden in Amsterdam, where Professor Ruysch has already sent me a packet"[90] – he initiated correspondence with Volckamer in the hopes of acquiring still more "exotic seeds." To this end, he enclosed with his very first letter a packet of seeds, in the hope that Volkamer would reciprocate (and at the very end of the letter he placed a want list of the plants he most desired).[91]

But Beringer soon proposed a means by which he might overcome the paucity of his resources in the botanical realm, and offer his patrician Nuremberg correspondent something in return that would put him on a more equal footing. After asking, at the end of the first letter, if Volkamer would "do him the friendship, of sending him any seeds he had left over," Beringer turned the topic of discussion in his next letter from plants to rocks: "and since I've also become a considerable connoisseur of minerals, shells, and other *naturalia*, I'd be happy to serve you in the future by sending you such things. . . ."[92] In this seemingly casual aside Beringer made up for his shortcomings as a *botanical* correspondent, by opening up a whole new realm of natural objects, namely that of minerals, where he might have more to offer.

In subsequent letters, Beringer cautiously began to follow through with his attempt to establish himself as a supplier of rocks, and to trade one natural kingdom for another. Shortly after the New Year in 1705, before the beginning of the growing season, he repeated his offer to Volkamer: "Perhaps I could serve you not only with some seeds I grew last summer, but also, along with several other curious people, with fossil rocks and other *naturalia*. . . ."[93] A month later, he referred to a "little box" he had sent and told Volkamer he had conveyed him one particular specimen as a token of his good faith: "I've undertaken to send you some *naturalia*, which happened to be available, that is to say in the package I mentioned before, I've included a piece of wood that gives off phosphorescence, found in the Lichtenberg mines not far from Erfurt, that was recently communicated to me via Mainz. I'd gladly have given you other fragments of minerals, fossils, and figured stones, except they were too heavy. . . ."[94] This token of sincerity delivered, Beringer felt free to turn back to the question of "what I'd also like to solicit" – namely foreign seeds.[95]

This pattern continued throughout the remainder of Beringer's correspondence with Volkamer. In exchange for the exotic seeds Volkamer supplied him with, Beringer sent him packet after packet of fossils, whether from Lüneburg,[96] the Harz,[97] or other German locales. Beringer sent seeds too,

[90] Beringer mss., letter of 14 March 1704.   [91] Beringer mss., letter of 14 March 1704.
[92] Beringer mss., letter of 17 March 1704.   [93] Beringer mss., letter of 10 Jan. 1705.
[94] Beringer mss., letter of 13 Feb. 1705.   [95] Beringer mss., letter of 13 Feb. 1705.
[96] Beringer mss., letter of 23 March 1705.
[97] Beringer mss., letters of 3 May 1705 and 23 December 1705.

of course, but often fewer than he would have liked, complaining repeatedly about the cold German winter and the occasional incompetence of his gardener. In short, the two naturalists worked out an arrangement agreeable to them both, one in which each felt as if he was contributing equally to the relationship. In this arrangement, each traded his special access to certain parts and products of the natural world. For Beringer, the natural worlds he had access to clearly included not just the fields and forests of Würzburg, but the contacts he had developed with other German naturalists: in other words, though his access was in many ways limited, it was far from being entirely "local." For Volkamer, on the other hand, his privileged access to Dutch sources of exotic seeds made him a desired correspondent for others, and enabled him to build his own networks further. It was the different spheres of the two naturalists that, if it did not make them a perfect match, at least kept their correspondence a mutually beneficial exchange.

Other naturalists' correspondence reveals many of the same themes. In each case, the naturalists involved, while basing their exchanges primarily on specimens that might be termed "local," and that they had deemed might be worthy of outside scrutiny, ended up exchanging far more than physical objects. The careers of a family of naturalists in Danzig (now Gdańsk), on the Baltic coast, illustrate this theme. Whereas the father, Jakob Breyne, had been a merchant, and had indulged in botanical pursuits during several visits to Amsterdam, his son Johann Philipp became a physician and, after his university years, chose to return to and remain in Danzig.[98] There he continued to exhibit his father's predilection for exotic plants, but also interested himself in local natural history, even going so far as to assist in the publication of a local flora of the area, from a pastor who had previously written a regional mineralogy.[99] Breyne's offers in his correspondence with such British natural-historical luminaries as James Petiver and John Woodward to send them "Things of Our Country" ended up netting him not only equivalent specimens in return, but the privilege of engaging in the ideals of "curious correspondence," as well as encouragement to work on

---

[98] As part of a broader investigation of the role of family and household in early modern natural inquiry, I am currently preparing a study on the Breyne family, including three of Johann Philipp Breyne's daughters, who painted many of the specimens he described. For more on the role of family and craft traditions in enabling the participation of women in science, see Londa Schiebinger, *The Mind Has No Sex? Women in the Origins of Modern Science* (Cambridge, MA: Harvard University Press, 1989) and Alix Cooper, "Homes and Households," in Katharine Park and Lorraine Daston, eds., *The Cambridge History of Science, Vol. 3: The Early Modern Period* (Cambridge: Cambridge University Press, 2006), 224–237.

[99] Georg Andreas Helwing, *Flora quasimodogenita: sive enumeratio aliquot plantarum indigenarum in Prussia* (Danzig: imprimebat Joannes Daniel Stollius, 1712) and *Lithographia Angerburgica* (Königsberg: literis Johannis Stelteri, 1717).

local natural-historical projects of his own.[100] In short, the exchange of local specimens not only connected naturalists separated from each other by the vast extent of the Holy Roman Empire, but also allowed them to engage in scholarly discussion of objects, manuscripts, books, and the whole apparatus of intellectual life at the time. In this way, dealing in the local could provide naturalists with legitimate access to privileged sources of natural-historical knowledge, while simultaneously enabling them to transcend sheerly "local" contexts.

Discussing a region's mineral wealth, then, was far from an affair focused merely on, and confined to, the region in question. Nor did it remain primarily focused, as it had previously, on precious metals alone. Instead, naturalists seeking to determine the mineral riches of their own territories, in the name both of self-sufficiency and of local pride, broadened their scope and looked *beyond* previous limits. They occupied themselves with the inventory of a wide range of minerals that were precious not so much in price, as in the significance that naturalists could assign to them. They read about the natural productions of other lands, and they hastened to inquire about acquiring specimens from these lands. They anxiously compared their own samples of local nature with these other examples of nature's differences elsewhere. It is within this context that the inventory of the mineral world came into being.

---

[100] Universitäts- und Forschungsbibliothek Erfurt-Gotha, Breyne Nachlaß, Chart. B 857b, fols. 42r and 38r. In a letter to Hans Sloane, Breyne confided that he was "now fully resolved to publish" his own local flora of Prussia, as soon as he got one more chance to travel through it (see British Library, Sloane ms. 4066, fol. 271); however, though he left many notes on his herborizing expeditions with his son, he never published the results of these expeditions.

4

The Nature of the Territory

Over the course of the seventeenth century, yet another alternative way of framing the study of local nature emerged in Europe, to take its place alongside the local flora and the regional mineralogy. This was the attempt to write the "natural history" of entire territories. This approach, most successfully formalized in England in the wake of Francis Bacon, was novel in several ways. One of the most important of these may be seen in the use of the term "natural history" itself. Unlike local floras and regional mineralogies, the new "natural histories" did not limit themselves merely to plants or rocks. Instead, they endeavored to capture the complete "natural history" of a place, its rich and varied complex of living and nonliving parts of nature, encompassing all of its varied kingdoms – including the animal kingdom, heretofore only rarely a subject for localist description.[1] Even more importantly, they moved beyond the local flora's close focus on a narrow radius around a town, claiming instead entire "counties" or "countries" as their fundamental unit. The narrowly defined project of the local flora was thus replaced by one far grander, and with a far more ambitious goal of natural description.

In the ways in which naturalists across Europe reacted to this development – or failed to do so – we can learn much about competing visions

---

[1] Descriptions of local fauna were relatively rare and unsystematic throughout most of the early modern period, consisting for the most part of treatises on less-common species near the margins of Europe, such as Scandinavian elk. Caspar Schwenckfeld, author of an early work on the plants and fossils of Silesia, did in fact author a so-called *Theriotropheum Silesiae* (Liegnitz: Impensis Davidis Alberti, 1603), but this must be seen as more of a general zoological work, given its inclusion of such animals as elephants (even though Schwenckfeld insisted an elephant had been seen once in Silesia, 89) and camels (which, Schwenckfeld admitted, *were* pretty hard to find in Silesia, 72). The eighteenth-century naturalist Georg Forster attempted to keep a record-book both of the flora and fauna he encountered in the region of Vilna, but within a year gave up on the latter, and started entering his botanical finds in the column for "fauna" as well as "flora"; see his "Diarium faunae floraeque Vilnensis," Universitätsbibliothek Göttingen, Hist nat 106, fol. 31. The central problem seems to have been that so many animal species were common to extremely wide areas of Europe. The study of local insects began to take off in the eighteenth century, though, in part through Martin Lister's pioneering work on English spiders, published in his *Historiae animalium Angliae* (London: Apud Joh. Martyn, 1678).

of natural inquiry in early modern Europe. For authors in different parts of Europe came, over the course of the seventeenth century, to develop different ideas about what "natural history" was, and how to write it. In Renaissance Europe, the term had not even been used very much. Though it would certainly have been familiar to anyone who had encountered one of the numerous printed editions of the ancient compiler Pliny's encyclopaedic *Natural History*; which was indeed quite popular, few naturalists had felt the need to term their own works "natural histories" along the Plinian model.[2] Rather, most naturalists had pursued projects somewhat more narrow than Pliny's. For example, while they might be willing to collect specimens widely from the three great kingdoms of nature (namely animal, vegetable, and mineral) for their cabinets, many of them tended in their writings to concentrate not so much on all three as rather on one or two of the specialist traditions targeting these kingdoms (namely zoology, botany, and mineralogy).[3] Early modern Europe was full of polymaths and virtuosi, but even they rarely cultivated *all* three traditions with equal intensity, let alone in the same work.[4] Lacking any pretensions to claiming their more limited inquiries as truly the equivalent of Pliny's, most writers discussing only a single natural kingdom in a work were therefore quite happy to regard that work as a contribution to the empirical genre of "history" more generally, labeling it (if *not* localist) a "history of plants" and so forth.[5] The local floras and regional mineralogies that proved so popular first in the seventeenth-century German territories, and then subsequently elsewhere in Europe may be seen likewise as falling within the more specialist traditions. Even though the authors of these repeatedly referred to their works as comprising "a part of natural history," their catalogues limited themselves to the inventory of individual

---

[2] Barbara Shapiro, "History and Natural History in Sixteenth- and Seventeenth-Century England: An Essay on the Relationship between Humanism and Science," in *English Scientific Virtuosi in the Sixteenth and Seventeenth Centuries*, ed. Barbara Shapiro and Robert G. Frank, Jr. (Los Angeles, CA: William Andrews Clark Memorial Library, 1979), 1–55.

[3] On the origins of these disciplines, see for example M. Möbius, "Wie sind die Bezeichnungen Zoologie, Botanik und Mineralogie entstanden?", *Jenaische Zeitschrift für Medizin und Naturwissenschaft* 77 (1944): 216–229, and W. Baron, "Gedanken über den ursprünglichen Sinn der Ausdrücke Botanik, Zoologie und Biologie," *Sudhoffs Archiv* 7 (1966): 1–10. On the full inclusion within early modern natural history of mineralogy, and its subsequent loss of popularity within natural history societies and the like, see D. E. Allen, "The Lost Limb: Geology and Natural History," in *Images of the Earth: Essays in the Environmental Sciences*, eds. L. J. Jordanova and Roy S. Porter (Chalfont St. Giles, UK: British Society for the History of Science, 1979), 200–212.

[4] On polymathy, see Anthony Grafton, "The World of the Polyhistors: Humanism and Encyclopedism," *Central European History* 18 (1985): 31–47 and Herbert Jaumann, "Was ist ein Polyhistor?", *Studia Leibnitiana* 22 (1990): 76–89.

[5] On history as laying the foundations for "learned empiricism" in Renaissance Europe, see the essays in Gianna Pomata and Nancy Siraisi, eds., *Historia: Empiricism and Erudition in Early Modern Europe* (Cambridge, MA: The MIT Press, 2005).

kingdoms of nature. Thus, while some works did appear labeling themselves "natural histories," they tended to be relatively rare. The idea of writing a book attempting to provide a complete "natural history," in the Plinian style, would have seemed daunting to say the least.

When, in the mid-seventeenth century, the project of writing complete "natural histories" finally did resurface, it did so most prominently in the British Isles, and in the context of the empiricism popularized by the English statesman Francis Bacon. Though Bacon's interests lay far more in theorizing the new science (in his *Novum Organum, Advancement of Learning* and elsewhere) than in actually practicing it, he did himself intend eventually to write a complete "natural history" which would provide the foundations for a reformed "natural philosophy"; his *Sylva sylvarum*, left unfinished at his death, had been intended to provide the first steps in this direction.[6] Though Bacon's program attracted some moderate interest on the Continent, it was his fellow Englishmen who ultimately did by far the most to popularize his method, especially after the founding in 1660 of the Royal Society of London. During the first decade of the Royal Society's existence, its members followed an ambitious and eclectic course, writing "natural histories" which explored topics far beyond the scope of the traditional three kingdoms, on subjects such as fire, air, nitre, and other assorted natural phenomena.[7] This reformulated idea of "natural history," in short, provided a key intellectual framework through which many natural inquirers in England learned to order their experiences.

Eventually, though, the English pursuit of "natural history" came to lose some of its original Baconian flexibility, and to focus more and more on taking inventory of the natural phenomena of specific places. In so doing it may be seen as following a trajectory – that of a move towards geographical organization – that the local flora, in particular, had already traced at the

---

[6] Francis Bacon, *Sylva Sylvarum* (London: Printed by J. H. for William Lee, 1627). See Paolo Rossi, *Francis Bacon: From Magic to Science* (London: Routledge & Kegan Paul, 1968); Julian Martin, *Francis Bacon, the State, and the Reform of Natural Philosophy* (Cambridge: Cambridge University Press, 1991); Antonio Perez-Ramos, *Francis Bacon's Ideal of Science and the Maker's Knowledge Tradition* (Oxford: Clarendon Press, 1988); and Paula Findlen, "Francis Bacon and the Reform of Natural History in the Seventeenth Century," in *History and the Disciplines in Early Modern Europe*, ed. Donald Kelley (Rochester, NY: University of Rochester Press, 1997), 239–260. Though Bacon's own program can be seen as sharing many similarities and continuities with other early modern forms of "historical" enquiry into nature, English-speaking writers since then have tended to grant him status as founder and inventor of virtually all empiricism; for a more nuanced approach, see Pomata and Siraisi.

[7] Though the group that would become the founding members of the Royal Society began to meet in 1660, they only received the charter for their project from Charles II in 1662; but the founding is conventionally assigned to the earlier date. The literature on the history of the Royal Society is vast; for the early years, see Michael Hunter, *Establishing the New Science: The Experience of the Early Royal Society* (Woodbridge, UK: The Boydell Press, 1989).

century's start. But this took new forms in mid-century, as scientific exchange and academy-foundation intensified, and virtuosi throughout Europe vied to establish themselves within newly-forming scientific communities. During the 1650s, following the English Civil War, the Hartlib circle in London had been instrumental in popularizing Baconian natural history, with the Dutch physician Gerald Boate's *Irelands Naturall History*, a work written in the wake of the Cromwellian reconquest of Ireland, as one result.[8] Similar efforts subsequently followed during the early years of the Royal Society, such as Joshua Childrey's *Britannia Baconica*, an attempt to survey the "natural rarities" of the remainder of the British Isles.[9] In the *Philosophical Transactions* of the Royal Society, numerous correspondents wrote in of various "particulars" of the natural history of their counties; indeed, such localist reports may be seen as constituting the greater part of the journal's output during much of the its early history.

Particularly influential in advancing this trend was Robert Boyle's "General Heads for a Natural History of a Countrey, Great or Small," which offered advice on this very topic: how to write the natural history of specific places. First published in the *Philosophical Transactions* in 1666, as a set of queries about specific phenomena he felt every "natural history" needed to address, this was eventually expanded and published in book form in 1692.[10] Though Boyle addressed his composition particularly to "Travellers

---

[8] Gerard Boate, *Irelands Naturall History* (London: for John Wright, 1652). On the Hartlib circle, see Charles Webster, *The Great Instauration: Science, Medicine and Reform 1626–1660* (New York: Holmes & Meier, 1973) and Mark Greengrass, Michael Leslie, and Timothy Raylor, eds., *Samuel Hartlib and Universal Reformation: Studies in Intellectual Communication* (Cambridge: Cambridge University Press, 1994). It is worth noting that during this period, one of the most active local naturalists was Sir Thomas Browne, who however was a royalist and thus would not have felt comfortable in the Hartlib circle. Browne was author not only of the famous works *Religio Medici* and *Pseudodoxia Epidemica*, but also of a local natural-historical manuscript, *Notes and Letters on the Natural History of Norfolk* (London: Jarrold & Sons, 1902); as this publication information suggests, he chose to circulate this in manuscript form, rather than submitting it to the press, and as a result it was first published only many centuries after his death.

[9] Joshua Childrey, *Britannia Baconica: or, The natural rarities of England, Scotland, & Wales* (London: printed for the Author, 1661); see also Christopher Merrett, *Pinax rerum naturalium Britannicarum, continens vegetabilia, animalia et fossilia in hac insula reperta inchoatus* (London: impensis Cave Pulleyn, typis F. & T. Warren, 1666).

[10] Robert Boyle, "General Heads for a Natural History of a Countrey, Great or Small," *Philosophical Transactions* 11 (April 1666): 186–189; and *General Heads for the Natural History of a Country, Great or Small; Drawn out for the Use of Travellers and Navigators* (London: printed for John Taylor, 1692). On the crucial role of early Royal Society projects in the origins of Boyle's "General Heads," and the broader significance of these kinds of "heads" or "queries" for early modern science, see Michael Hunter, "Robert Boyle and the Early Royal Society: A Reciprocal Exchange in the Making of Baconian Science," *British Journal for the History of Science*, forthcoming. The kinds of information requested in these

and Navigators," in the hope of steering them into generating rigorously scientific travel accounts, his model was in effect a universal one, which could be applied to a wide range of "countries," both at home and abroad. In this and other prescriptive treatises,[11] and in the numerous natural histories subsequently written of specific places, we see a growing (though relatively tacit) consensus that the "natural history" of individual places provided, self-evidently, a fundamental means for organizing knowledge. What such places might consist of was by no means settled; authors mingled terms such as "country," "county," "region," and "territory" seemingly interchangeably. During the 1660s, however, the Baconian "natural history" of the territory remained largely a British phenomenon; authors elsewhere in Europe were not yet convinced of its usefulness.

This chapter will argue that the late seventeenth and early eighteenth centuries saw a series of powerful tensions emerge between alternative visions of natural history, centered on differing concepts of "nature" and "territory." One such vision, for example, was articulated by Henry Oldenburg, secretary to the Royal Society of London and a transplanted German, who attempted in his letters to interest his correspondents in writing Baconian "natural histories" of their regions. Oldenburg had considerable success in eliciting such documents from English and colonial recipients of his pleas. However, his attempts to persuade his colleagues on the Continent to take part in writing natural histories according to the new pattern must, as we shall see, be judged a miserable failure. While Oldenburg's efforts here came to no fruit, though, those of the Swiss entrepreneur and polymath Johann Jakob Scheuchzer were to succeed, through his prolific natural histories of Switzerland, in interesting continental readers in conducting natural histories of their territories. As entrepreneurs in assorted German territories took up Scheuchzer's goals and methods and marketed them at court, though, they generated what must be seen as yet another style of regional natural history, one specifically adapted to central European administrative and cameralist concerns. By following the trajectory of the natural history of the territory

queries can be seen, in fact, as representing a systematization of the kinds of information already present in many colonial natural histories, as well as in Plinian natural history itself.

[11] See also John Woodward, *Brief Instructions For Making Observations in all Parts of the World* (London: printed for Richard Wilkin, 1696), for a subsequent example. For the use of such queries as means of gathering information, see Justin Stagl, "Vom Dialog zum Fragebogen: Miszellen zur Geschichte der Umfrage," *Kölner Zeitschrift für Soziologie und Sozialpsychologie* 31 (1979): 611–638, and also his *A History of Curiosity: The Theory of Travel 1550–1800* (Chur, Switzerland: Harwood Academic Publishers, 1995), *passim*. As Stagl shows, the use of questionnaires was not confined to England or the Royal Society, but became increasingly popular throughout Europe over the course of the early modern period.

during this period, we may simultaneously trace shifting ideas of nature and territory, as nature came increasingly to be seen as a stable entity, and territory to be seen as the embodiment of politics in nature.

"PROCURE US AN ACCOUNT"

Sitting at his writing-desk in 1660s London, at least one man wondered about the lack of non-English "natural histories." This was Henry Oldenburg, a German by birth (from the north German town of the same name) who, during the aftermath of the Thirty Years' War and the late years of Cromwell's rule, had taken up residence in London. There he joined fellow German expatriates Samuel Hartlib and Theodore Haak in their attempts to reform scientific and intellectual life.[12] In the process, he began to build up an extensive correspondence network with scholars and other letter-writers on the Continent, one that would ultimately extend throughout Europe. After the end of the Commonwealth, Oldenburg found a way to continue his activities as an "intelligencer," by affiliating himself with the newly-founded Royal Society of London. By dint of his Continental connections, he was soon appointed the Society's secretary and, as a logical extension, the editor of its *Philosophical Transactions*.[13] Brokering foreign news could be dangerous; on more than one occasion, Oldenburg found himself under suspicion of espionage.[14] Yet throughout the decades of the 1660s and 1670s, Oldenburg used his status as the Royal Society's chief correspondent with foreign scholars to make repeated inquiries as to the "natural history" of areas around the world. In particular, he urged his correspondents time and time again to contribute "natural histories" of their own territories. By observing Oldenburg's attempts to promote an English Baconian style of natural history elsewhere in Europe, and his overwhelming lack of success in this endeavor, we may begin to see some of the conflicts and tensions that surrounded the enterprise of natural history in early modern Europe, and the different "local" styles in which that enterprise could be framed.

In his correspondence, Oldenburg may be seen as pursuing the Royal Society's general aim of collecting empirical knowledge. For Oldenburg, this entailed most often the seeking of knowledge about the natural history of

---

[12] For more on the "three foreigners," see H. R. Trevor-Roper, "Three Foreigners: The Philosophers of the Puritan Revolution," in *Religion, the Reformation, and Social Change and Other Essays* (London: Secker and Warburg, 1984), 237–293; Dorothy Stimson, "Hartlib, Haak and Oldenburg: Intelligencers," *Isis* 31 (1939–1940): 309–326; and Webster, *The Great Instauration, passim.*

[13] On Oldenburg's role as intelligencer for the Royal Society, see Marie Boas Hall, *Henry Oldenburg: Shaping the Royal Society* (Oxford: Oxford University Press, 2002).

[14] See Douglas McKie, "The Arrest and Imprisonment of Henry Oldenburg," *Notes and Records of the Royal Society of London* 6 (1948): 28–47.

particular *places*. Oldenburg's strategy as a correspondent was to encourage the "advancement of learning" both in Britain and abroad by suggesting to his readers specific ways in which they, in their local situations, might contribute to the Royal Society's broader agenda. Oldenburg thus frequently made quite specific requests to correspondents about ways in which they might make themselves especially useful. For many of his foreign correspondents, these requests concerned the "natural history" of the regions they inhabited. Seeking to gather empirical knowledge from all corners of the world, Oldenburg in effect broke that knowledge down geographically, into far larger units than those the local flora had covered; he often urged correspondents to report on entire territories. Such requests may have seemed reasonable to Oldenburg, but many of his correspondents were to shy away from the ambitious conceptual and geographical scope of the projects he proposed.

Oldenburg's correspondence with Hevelius, the famous astronomer living in the Baltic port of Danzig in far-away Prussia (now Poland), reveals his considerable interest in natural history and his *modus operandi* for soliciting information about the natural history of particular territories, while also suggesting some reasons for the reluctance of his correspondents to do as he asked. Oldenburg wrote many letters to Hevelius in Latin, the language of international scholarly communication, rather than that of their birth. These letters were hardly confined to the astronomical matters Hevelius had built his observatory to track,[15] but were instead liberally sprinkled with queries about the natural history of Prussia. Even before Oldenburg first contacted Hevelius, we can follow the direction of his thinking in a letter he wrote to Boyle on January 27, 1665/6:

We have thoughts of engaging as many of ye Society, as are cordiall and have opportunity, to observe and bring in, what is any wayes considerable of Naturall productions in England, Ireland, Scotland; everyone his Symbol, for ye bringing together a Naturall History of what is in yeso kingdoms, as well as we intend to collect what is abroad, by enlarging our Correspondencies every where, we can.[16]

As this passage reveals, Oldenburg saw his responsibilities at the Royal Society as including the collecting of natural-historical information not only internal to the "kingdoms" of the British Isles, but in foreign territories as well. To organize this information, he proposed to use the frameworks provided by two of the Society's chief movers, Hooke and Boyle. "Mr. Hook has also ready (having shewed it me and others) a Method for writing a

[15] For an account of Hevelius's interactions with the Royal Society that *does* focus on the astronomical affairs, see Steven Shapin, *A Social History of Truth: Civility and Science in Seventeenth-Century England* (Chicago, IL: University of Chicago Press, 1994), 272–308.

[16] Rupert Hall and Marie Boas Hall, eds., *The Correspondence of Henry Oldenburg* (Madison: University of Wisconsin Press, 1965–1986), III, 32 (hereafter cited as "Oldenburg"). The Halls' translations will be used unless otherwise noted.

Naturall History, wch, I think, cuts out work enough for all Naturalists in ye World..." [17] About a month later, Oldenburg reported to Boyle that

I intend, within a few dayes, to write to Hevelius, and to engage him to give or procure us an Account of ye way of making Potashes, and of ye Salgemmae-mines in Poland. If any thing else come in yr mind, worthy to be inquired after in Borussia, Poland, Liefland, I pray, send it to me, and I shall diligently recommend it. Some Enquiries about Amber would not be amisse for those parts. [18]

Here we see Oldenburg pondering in advance which queries to send Hevelius – not only questions about specific natural objects, such as potash and amber, but also requests for information on particular territories. Hevelius, Oldenburg clearly hoped, might be willing to serve as a correspondent to the Royal Society on matters of the natural history not only of Prussia but also of neighboring Baltic areas. Even though he knew that Hevelius's talents and interests lay more along astronomical lines, Oldenburg did not see natural history as a specialist discipline, as something capable of being pursued only by trained physicians, or by specialists in either plant or mineral realms. Rather, he thought it the duty of every virtuoso or would-be member of the scientific community to contribute to the establishment of natural-historical knowledge, whether by his own efforts or by gathering information from others.

Oldenburg's letter to Hevelius on March 30, 1666 thus posed questions he had been turning over in his mind for some time.

Here is your list of those who are citizens of our scientific community [i.e. members of the Royal Society]....And as they are first of all engaged upon natural history, and to that end are busy in seeking out in all parts those things with which Nature enriches different regions, we trust you to add your contribution. For this purpose, you see the annexed queries; we earnestly beg you either to prepare answers to these yourself or to procure them from friends with whom, perhaps, you correspond in Borussia, Poland, Sweden, and Muscovy. [19]

Through the list of queries he provided, a procedure which was to become extremely common in soliciting information for the metropolis from the provinces, we can see Oldenburg as attempting to direct Hevelius's efforts towards subjects deemed of particular interest in London. Here we see also, however, that Oldenburg has in this letter significantly extended the areas for which Hevelius would be held responsible, adding not only Sweden, but all of Muscovy, i.e. the Russian empire. Oldenburg was thus requesting information on a scale far beyond the ability of Hevelius, or any other individual Baltic resident, to fulfill. Hevelius's response of several months later (June 23, 1666) was a bit too distressed about an astronomical controversy to pay much attention to natural history, but in his last paragraph, he declared that

[17] Oldenburg, III, 32.   [18] Oldenburg, III, 47.   [19] Oldenburg, III, 73, 76.

he would work on the problem, and that he had written to friends about Oldenburg's queries.[20]

These initial exchanges set the pattern for the remainder of Oldenburg's correspondence with Hevelius. Oldenburg repeatedly prodded Hevelius to send him answers to his queries, hopefully maintaining, for example, in a letter of August 24, 1666 that "meanwhile, we are very glad that you are so far forward in making your contribution towards preparing a Natural History."[21] Several weeks later, he anxiously added: "meanwhile you will not forget those points I commended to you, concerning Natural History, certain matters in physics, and the procurement of a literary correspondence in your northern clime."[22] Hevelius's responses to Oldenburg, meanwhile, show a certain amount of guilt at his delays in responding adequately to the Royal Society's requests. In a letter of October 19, he made a brief attempt at answering the queries, including one paragraph full of answers; but he largely ducked the task, by claiming that he had passed on the queries to others of his acquaintance. He particularly regretted his ignorance on the subject of amber; Hevelius claimed, however, that he was looking forward to a trip to Königsberg, where he hoped he would find out more about this valuable Baltic commodity.[23]

But Oldenburg was scarcely satisfied by Hevelius's answers. This may be seen in the somewhat gentle rebuke he delivered to Hevelius in his letter of February 27, 1666/7: "The answers to some of our questions returned both by yourself and some of your friends are apt and acute; we greatly wish that the remainder may be satisfied in the same sincere manner. This year's weather has been most fit for so doing; and we have no doubt but that your friends will, at your request, apply themselves with all their might to the outstanding ones."[24] Half a year later, with the approach of fall, Oldenburg wrote again to Hevelius, wondering why he had as yet received no reply. "Winter now approaches, which is the season for urging ingenious men everywhere in your parts to strive after replies to the rest of our questions about gold, amber, and potashes and the way of preparing them, based upon careful observation and experiment."[25] At this point Oldenburg's correspondence with Hevelius ceases, and it is unclear whether Oldenburg ever received the information he desired about Baltic natural history. Clearly, Hevelius was not a satisfactory correspondent in these matters; whether this may have been because he was suspicious that Oldenburg's prying queries verged on espionage, or simply preoccupied with his own astronomical activities, we may never know. Clearly, though, Oldenburg had failed to awaken in him an interest in the natural history of his region, let alone the commitment to pursue this on a larger scale and/or to communicate it to a broader scholarly audience.

[20] Oldenburg, III, 170, 172.   [21] Oldenburg, III, 216, 219.   [22] Oldenburg, III, 223–224.
[23] Oldenburg, III, 248–249, 255.   [24] Oldenburg, III, 350, 353.   [25] Oldenburg, III, 515–516.

Oldenburg's interest in gathering information about the natural history of specific places appears in numerous letters to other correspondents. In a letter of April 10, 1666 which he sent to an English gentleman in Italy, Oldenburg urged him to find someone to write a natural history of the peninsula.[26] In a later letter to Italy, addressed on June 30, 1669 to George Cotton, an English Jesuit in Rome, he again begged his reader to "sollicite some ingenious and curious men over all Italy, to write the Naturall History of that Excellent Countrey...."[27] Likewise, in a set of "Inquiries for Germany and Hungary/ Commended to Mr. Thomas Coxe" on June 30, 1669, he concluded, "let men who are notably ingenious and industrious be urged to compile the natural histories of the regions wherein they dwell with the utmost care and reliability."[28] That this geographically-oriented approach formed the basis for Oldenburg's efforts to acquire information from home as well as abroad may be seen, for example, in a letter of his to a certain Towneley on March 2, 1668/9 in which he "desired him, to send us ye Observables of his Contry and the circumjacent parts" (the "country" in question being the English county of Lancashire, clearly in Oldenburg's eyes yet another territory to be canvassed for *naturalia*).[29]

Perhaps owing to his own German origins, Oldenburg made repeated efforts to elicit the natural histories of the German territories in particular. One of the most promising responses he received was that given by the Hamburg bookseller and physician Martin Vogel, in a letter of January 31, 1671/2. In reply to Oldenburg's queries about the Harz forest, Vogel reported that he was too far away to research the area himself, and that the literature on the topic (for example, Thal's local flora written almost a century previously) was inadequate: "Long ago Johann Thal, a physician of Nordhausen, proved very diligent in collecting plants from this forest in a single edition. The history of metallic bodies is chiefly to be sought in Lonicer and others.

---

[26] Oldenburg, III, 87; see also III, 618. "One thing more there is, I must sollicite; vid. yt it being a part of ye R. Society's dessein, to compose a good Nat. History, to superstruct, in time, a solid and usefull Philosophy upon, and ye compiling of ye natural Histories of particular Contries appearing very conducive to such a dessein, you would use your interest in Italy to excite some able and diligent persons to set upon yt work for ty contry; as I hope, the like will be done, by our importunity, in Spaine, Portugall, France, Germany, Poland, Hungary etc. And it being, of no slight importance, to be furnisht wth pertinent Heads, for ye direction of inquirers, there hath been printed in ye Last of the abovemention'd Phil. Transactions, a List of Generall and comprehensive Articles for yt purpose, of wch I shall endeavor to get some [copies] transmitted to you...." The last-mentioned piece is almost certainly Boyle's "General Heads for the Natural History of a Countrey, Great or Small."

[27] Oldenburg, VI, 79. No natural history of Italy along English lines was to be forthcoming for quite some while; however, at least two French local natural histories did end up being written, namely Jean-Scholastique Pitton's *De conscribenda historia rerum naturalium Provinciae* (Aix: Apud Carolum David Regis, 1672), and, many years later, Jean Astruc's *Mémoires pour l'histoire naturelle de la province de Languedoc* (Paris: Chez Guillaume Cavelier, 1740).

[28] Oldenburg, VI, 76–77.   [29] Oldenburg, V, 427.

We lack an account of the animals."[30] Vogel even suggested that he himself
was not uninterested in the project of writing about the natural history of
the Harz, and would gladly be willing to do such a thing, if only there were
sufficient financial support:

> If I could look forward to some royal stipend, I would seek out again places I formerly
> examined in that forest and write a History from which perhaps nothing would be
> found lacking by lovers of such things. Now I am busy attending the sick, in order
> to nourish my family properly. Nevertheless, I can publish a natural history of the
> forest, such as Childrey gave of England, if I combine my extracts and observations
> with Thal's catalogue, the excerpts from Lonicer, and so forth. Would that I had the
> help of colleagues![31]

As a physician, Vogel pointed out, his primary source of income was from
his patients, who all presumably lived in Hamburg or its environs and thus
confined him to this locality as well. The problem of patronage was thus
a significant one for Vogel, who like many other continental Europeans
seems to have envied the "royal stipends" and support which English scientists were (wrongly) thought to receive from their royal connection with
Charles II.

Yet it is also significant that Vogel possessed and cited an English book,
namely Childrey's *Britannia Baconica*, as a model.[32] Childrey's natural history of Britain seems to have been helpful to Vogel in conveying an image of
what Oldenburg actually wanted from him. Vogel's vision of a complete "natural history of the forest" represented an important departure from models
that previously existed in the German territories, such as the local flora and
regional mineralogies with their specialist forms. In the absence of patronage, such an ambitious project could not be undertaken. Vogel noted, though,
that quite a few other learned men in the German territories *were* interested
in natural-historical topics. He therefore suggested that Oldenburg contact
these other literati, helpfully providing many names: "You should excite Mei-
bom, an inquisitive and learned man, professor at Helmstadt; Bernhardides,
physician at Wolfenbüttel but formerly practicing medicine at Zellerfeld in
the Thuringian forest; Stochus, physician at Goslar...; Ramlovius, physician at Ostlerode, and Gryling, physician at Stolberg."[33] It is noteworthy
that almost all of the men Vogel named were medical men, who, with their
university training in "physic," were seen as suitable to pronounce upon
natural things. The problem with Oldenburg's request, then, was *not* that
the German territories lacked naturalists – the sheer number of local floras and similar works discussed in the previous chapters suggests rather the
opposite – but instead that natural history was conducted in a very different
style than that envisioned by Oldenburg. Namely, it was conducted in the

[30] Oldenburg, VIII, 513, 515.   [31] Oldenburg, VIII, 513, 515.
[32] See Childrey, discussed above.   [33] Oldenburg, VIII, 513, 515.

style of the local flora and the regional mineralogy; and the German authors of these latter works were busy physicians with little spare time to conduct ambitious new projects, rather than the eclectic mix of leisured English gentry and active travelers the Royal Society had tapped, a great number of whom *had* proven extremely willing to respond to Oldenburg's queries. In short, a very different set of conditions existed in the German territories than in England, and concerns about patronage only begin to explain the extent of these differences.

As a substitute for a work of his own, Vogel wrote to Oldenburg in one of his next letters on July 25, 1671 that he would send him a copy of a recently published "Description of Fossils in the Region of Hildesheim," in other words, a regional mineralogy.[34] Oldenburg, however, was clearly not satisfied with this specialist work. He thus went on to beg Vogel in a letter of November 14, "See to it, please, that the deepest recesses of the Harz mountains (among many other places) are investigated by skilled persons with an eye for their minerals, plants, and animals, and let their findings be brought together into a proper narrative."[35] Here Oldenburg did not respond to Vogel's previous concerns, but merely reiterated his own vision of the kind of natural history that he wanted, one which Vogel had already made clear was foreign to his own needs and situation. In the circumstances, it is not surprising that Vogel simply let the matter drop.

These kinds of difficulties were certainly not unique to the German territories. A letter written to Oldenburg by Justel, one of his French correspondents, reported on an interesting piece of regional news:

He who gave me the drawing of the roads of Savoy assured me that an individual is working on the history of the Alps which are adjacent to Savoy and the neighborhood of the country subject to the duke of Savoy. He is taking very great care and many pains to describe all the places through which he passes, having intelligent people for that. He will note the minerals, crystals, animals, plants, and flowers to be found there; thus the work is worthy of curiosity and everyone must hope that it will be finished. It is feared lest the death of the duke will cause this honest man to cease work, no one wishing to work uselessly and without some particular hope when one must make disbursements.[36]

This example presents almost a textbook case of the perils of depending on personal patronage. In the German context, for example, physicians compiling local floras in the specialist tradition had at least the reasonable assurance that universities, and their medical faculties, might continue to support the publication of works even after the death of their initial compilers.

---

[34] This would almost certainly have been Friedrich Lachmund, *Oryktographia Hildesheimensis, sive admirandorum fossilium, quae in tractu Hildesheimensi reperiuntur* (Hildesheim: sumptibus autoris, Typis viduae Jacobi Mülleri, 1669). This work is discussed in the previous chapter.

[35] Oldenburg, VIII, 356–357. [36] Oldenburg, XI, 402–403.

Indeed a fair proportion of local floras were published posthumously, by their authors' sons or colleagues. In this way, although the production of local floras remained by all accounts a time-consuming and sometimes even multigenerational matter, the university setting minimized the necessity for external support. Princely courts, where patronage was based much more on personal relations with a ruler, offered no such guarantee of continuity. Nor were Oldenburg and his colleagues able to offer much assistance in this regard. The style of natural history Oldenburg proposed simply did not seem at all feasible to many of his Continental correspondents.

In the late 1670s, the publication of Robert Plot's *Natural History of Oxford-shire* offered Oldenburg a new tool with which to convince others to write "natural history" in the English style.[37] Oldenburg was repeatedly to mention this work to correspondents, suggesting that it offered a perfect model for how the natural history of a territory ought to be written. Plot's book, a large and compendious volume containing his researches not only on natural history but also on the antiquities of his county, certainly inspired other English authors of what came to be known there as "county natural histories." Plot himself, for example, wrote a subsequent *Natural History of Stafford-shire*, while others became active in contributing notes on their own counties to the *Philosophical Transactions*, and eventually published entire natural histories of their own.[38] Plot's model, then, of natural history mixed with antiquarianism, was clearly effective in English contexts.[39] As

---

37 Robert Plot, *The Natural History of Oxford-shire, Being an Essay toward the Natural History of England* (Oxford: Printed at the Theater, 1677). On Plot himself, see Stan A. E. Mendyk, "Robert Plot: Britain's 'Genial Father of County Natural Histories,'" *Notes and Records of the Royal Society of London* 39 (1985): 159–177; on his broader context within English writing, see Lesley B. Cormack, "'Good Fences Make Good Neighbors': Geography as Self-Definition in Early Modern England," *Isis* 82 (1991): 639–661; Stan A. E. Mendyk, *'Speculum Britanniae': Regional Study, Antiquarianism, and Science in Britain to 1700* (Toronto: University of Toronto Press, 1989); and Kenneth Robert Olwig, *Landscape, Nature, and the Body Politic: From Britain's Renaissance to America's New World* (Madison: University of Wisconsin Press, 2002). See also Charles Withers, *Geography, Science, and National Identity: Scotland since 1520* (Cambridge: Cambridge University Press, 2001).

38 Robert Plot, *The Natural History of Stafford-shire* (Oxford: Printed at the Theater, 1686). For examples, see Charles Leigh, *The Natural History of Lancashire, Cheshire, and the Peak, in Derbyshire: with an Account of the British, Phoenician, Armenian, Gr. and Rom. Antiquities in those Parts* (Oxford: Printed for the Author, 1700); Thomas Robinson, *An Essay towards a Natural History of Westmorland and Cumberland* (London: by W. Freeman, 1708); and John Morton, *The Natural History of Northampton-shire; with Some Account of the Antiquities. To which is Annex'd a Transcript of Doomsday-Book, so far as it relates to That County* (London: Printed for R. Knaplock . . . and R. Wilkin, 1712). It is worth noting that of these authors, only Leigh was a physician; the others were clerics with antiquarian interests.

39 Another example of how English naturalists tended to combine aspects of antiquarianism with their natural history can be seen in the case of the famous botanist John Ray himself, who collected in print not only local vernacular plants but local vernacular proverbs; see his *A Collection of English Proverbs* (Printed by J. Hayes, 1670). On this project, see Jo Gladstone, "'New World of English Words': John Ray, FRS, the Dialect Protagonist, in the

we shall see, though, it was to prove less so in Continental ones. In a letter to Hevelius on June 13, 1677, Oldenburg used this new resource to attempt to persuade Hevelius, whom he had first solicited for information on amber over a decade and a half earlier, to contribute further to the knowledge of German natural history: "You, best of men, will solicit the industrious and inquisitive men of your Prussia to try and tackle something of the sort [i.e. like Plot's book] so that by compiling a reliable natural history of your region they may make their contribution to the design successfully begun here."[40] Hevelius, alas, remained as unforthcoming as before.

But Oldenburg tried again, as always. In a letter to the famous philosopher Leibniz, Oldenburg lauded Plot as an example to be followed, and pleaded with him to undertake a natural history of Lüneburg and Braunschweig (anglicized as "Brunswick"):

A certain Oxford man, named Doctor Plot, has lately sent into the world a natural history of Oxfordshire, as a specimen of a plan he has undertaken of compiling the natural histories of all the English counties. In this Oxfordshire history he has noted and recorded everything which he could have observed himself, with the aid of many skillful gentlemen, concerning the nature, arts and antiquity of that county.... I have already written to many of my friends overseas of what has been accomplished in this matter by us here and have urged them to follow this example and attempt something similar in their own corners of the globe and in this way be pleased each to make his own contribution to building the structure of a universal history of nature.[41]

It is worth noting that Oldenburg described Plot's *Natural History* as a study not only of "nature" but of "arts and antiquity" in Oxfordshire. To Leibniz, who spent decades on researching the history of his princely patrons' families, this hint of antiquarianism would not necessarily have been uncongenial.[42] Furthermore, Oldenburg phrased his overall effort as one that would unite scholars all over the globe, in the process appealing to Leibniz's passion for schemes of intellectual communication.[43] The problems emerge in the second half of Oldenburg's plea:

I am quite confident that you, famous Sir, will not lag behind the foremost, but will devote your energies to have a similar history compiled of the broad regions under the sway of their Serene Highnesses of Lüneburg and Brunswick, to which I am very confident those most wise and learned as well as warlike heroes will firmly

Context of his Times (1658–1691)," in *Language, Self, and Society: A Social History of Language*, ed. Peter Burke and Roy Porter (Cambridge: Polity Press, 1991), 115–153. In contrast, German local florists had tended to pursue antiquarian researches only into the history of medicine itself, and of their university or university town.

[40] Oldenburg, XIII, 289–290.   [41] Oldenburg, XIII, 320, 323.

[42] For Leibniz's own antiquarian researches, see Günter Scheel, "Leibniz und die deutsche Geschichtswissenschaft um 1700," in *Historische Forschung im 18. Jahrhundert*, ed. Karl Hammer and Jürgen Voss (Bonn: Ludwig Röhrscheid Verlag, 1976), 82–101.

[43] See Ayval Ramati, "Harmony at a Distance: Leibniz's Scientific Academies," *Isis* 87 (1996): 430–452.

and generously apply their authority and a share of their highest powers. No doubt many things worthy of remark are to be found in the Harz mountains, a good part of which is ruled by those meritorious princes. It would surely be much to be regretted if all those things should be forever concealed from the knowledge of philosophers, and never brought to light so that they may form a worthy part of nature's store.[44]

Here Oldenburg attempted to argue that Leibniz's princely patrons would look with favor on a naturalistic equivalent of the antiquarian project he was currently involved in. By mentioning the Harz, however, only part of which was under the control of the house in question, Oldenburg weakened his case. At any rate, Leibniz continued his genealogical researches, and the projected natural history never saw the light of day.

Perhaps the most revealing glimpse into the reasons for Oldenburg's lack of success with his Continental correspondents, and those in the German territories in particular, comes in a letter to Oldenburg dated January 12, 1664/5. Here, Philipp Jakob Sachs, a German small-town physician and the president of the "Academia Naturae Curiosorum," a recently-founded German medical society made up of learned physicians from all over the Holy Roman Empire, paid his respects to Oldenburg and the Royal Society.[45] This letter is often cited by historians of early modern scientific societies to explain why the Holy Roman Empire did not develop "strong" scientific institutions like the British and French; but it is equally revealing of broader differences in scientific style. For in this letter may be seen clear signs of a clash between science as practiced in the German territories, and as practiced in London. Sachs, for example, praised the Royal Society's determination to do experiments "in your praiseworthy English way," unobtrusively implying that Germans might, perhaps, do experiments differently.[46] He then went on to provide a comparison of the two learned societies, modestly stressing the shortcomings of his own:

Even Germany felt a few years ago the same incentive towards founding a certain college, under the name of the Academy of the Investigators of Nature. But just as bodies stretched out between widely separated nodes are a weaker structure than those that are shorter and more compact, so our college dispersed over the broad provinces of Germany is of less strength than the Illustrious Society with its permanency in London.[47]

---

[44] Oldenburg, XIII, 320, 323.
[45] For more on the Academia, see Mason Barnett, "Medical Authority and Princely Patronage: The *Academia Naturae Curiosorum*, 1652–1693" (Ph.D. thesis, University of North Carolina, Chapel Hill, 1995); R. J. W. Evans, "Learned Societies in Germany in the Seventeenth Century," *European Studies Review* 7 (1977): 129–151; and Erwin Reichenbach and Georg Uschmann, eds., *Nunquam otiosus. Beiträge zur Geschichte der Präsidenten der Deutsche Akademie der Naturforscher Leopoldina* (Leipzig: Barth, 1970).
[46] Oldenburg, II, 342, 345.   [47] Oldenburg, II, 342–343, 345.

Emphasizing the decentralized nature of intellectual life in the Holy Roman Empire, Sachs gave a further geographic explanation for what he saw as his Academy's modest achievements:

Seafaring leaves all corners of the earth open to you. We Germans know only narrower limits, magnates with slenderer purses, and a shortage of grand patrons. Our college is scattered hither and yon; its members are medical men exhausted by the cares of practice who find few spare hours for natural experiments; and so I say nothing about the frequent catastrophes to the letters we exchange, even on the roads, or about the wars whose tumults plague now this province, now that, disrupting the peace of the muses.[48]

Sachs thus blamed German deficits in natural-historical productivity not only on the constraints of patronage and profession, but on the more fundamental set of structural problems created by life in the fragmented German territories. This fragmentation had, of course, been little hindrance to the production of local floras, which had continued to be written even during the ravages of the Thirty Years' War. But Sachs was concerned that German experiences, during the past few disruptive decades, of being cut off from broader European intellectual contact had doomed him and his fellows to the status of provincials. He enclosed with his letter a copy of a highly baroque treatise of his, with the title *Oceanus Macro-Microcosmicus*, but worried that it might be found out of date, or at the very least not written in the English plain style: "I can readily foresee that these things will not be welcome to palates accustomed to more delicate foods."[49] Following this invocation of what might be termed scientific as well as literary "taste," Sachs concluded his letter with the comment that "the progress of our college has been hindered in many ways, but we can hope for better luck in the future, if a flood of learning and experimental truth can be brought through epistolary channels from your British ocean to fertilize our more sterile Germany."[50] With this, Sachs ended the discussion. But the implication was clear: such significant differences in scientific styles separated the two regions that in no way could they be easily overcome.

TRANSLATING THE NEW SCIENCE

Ultimately, the natural history of the territory was to make its way into continental Europe, not via Oldenburg and the Royal Society, but rather through the inspiration and model of a Swiss physician. Johann Jakob Scheuchzer

---

[48] Oldenburg, II, 343, 345.

[49] Oldenburg, II, 343, 346. The book in question was Philipp Jakob Sachs von Lewenhaimb, *Oceanus Macro-Microcosmicus* (Breslau: Sumtibus Esaiae Fellgiebelii, 1664).

[50] Oldenburg, II, 343, 346. This statement may be seen as rather ironic, given the way in which local floras and mineralogies had unfailingly stressed the "fertility" of individual regions of Germany.

was a recent medical school graduate when he returned to Zürich from his studies abroad and penned the first of a set of queries about the natural history of Switzerland. Over the course of the next several decades, and many excursions into the Alps and all over Switzerland, Scheuchzer was to devote the majority of his energies towards assembling materials for a "natural history of Switzerland," with the ultimate goal of collecting information about all possible aspects of Switzerland's natural world, viewed from every angle.[51] Scheuchzer is perhaps best known today as the author of a lavishly-illustrated natural-theological *Physica Sacra*, or "sacred physics," and of two other striking works, one titled the *Piscium querelae et vindiciae*, or "complaints and claims of the fishes" (1708), which argued that fossils were the remains of organic beings, and presented the fish themselves as making this case; and the other titled *Homo diluvii testis et theoskopos*, which (mistakenly) argued that a fossil bone was of human origin, showing that humans too had perished during the Noachian Flood.[52]

The bulk of Scheuchzer's massive output, however, concerned itself with the natural history of Switzerland in one way or another. In 1706, not long after printing his initial queries, he started publishing brief narratives on remarkable Swiss natural phenomena in the, for naturalists, highly unusual format of a weekly magazine; by 1716, he had begun to assemble these fragments into a single thick volume, which he published in several installments as the *Natur-Historie des Schweitzerlandes*.[53] Most of these works were published in German as well as Latin, thereby reaching an audience both near and far. Bridging the gap between the Royal Society's quest for "natural history" and his own intimate acquaintance with Central European regionalism, Scheuchzer's model of how to write the natural history of a territory came to attract numerous followers in the early years of the eighteenth century. Integrating a compelling set of arguments for the usefulness of a

---

[51] Note that although Johann Jakob Scheuchzer had a brother with a strikingly similar name, Johann Scheuchzer, who also wrote on the natural history of Switzerland, the two should not be confused. Johann Scheuchzer is best known for his study of Swiss grasses, *Agrostographiae Helveticae* (Zürich: sumptibus autoris, 1708).

[52] Johann Jakob Scheuchzer, *Kupfer-Bibel* (Augsburg & Ulm: gedruckt bey Christian Ulrich Wagner, 1731–1735); Johann Jakob Scheuchzer, *Piscium querelae et vindiciae* (Zürich: sumtibus Authoris, typis Gessnerianis, 1708); Johann Jakob Scheuchzer, *Homo diluvii testis et theoskopos* (Zürich: typis Joh. Henrici Byrgklini, 1726). For more on these works, see Melvin E. Jahn, "Some Notes on Dr. Johann Jakob Scheuchzer and on *Homo diluvii testis*," in *Toward a History of Geology*, ed. Cecil J. Schneer (Cambridge, MA: The MIT Press, 1969), 192–213, and more generally Michael Kempe, *Wissenschaft, Theologie, Aufklarung: Johann Jakob Scheuchzer (1672–1733) und die Sintfluttheorie* (Epfendorf: Bibliotheca Academica, 2003).

[53] Johann Jakob Scheuchzer, *Beschreibung der Natur-Geschichten des Schweizerlandes* (Zürich: In Verlegung des Authoris, 1706–8); Johann Jakob Scheuchzer, *Helvetia historia naturalis, oder Natur-Historie des Schweizerlandes* (Zürich: in der Bodmerischen Truckerey, 1716–1718).

broadened "natural history," Scheuchzer spoke a symbolic language continental Europeans could understand.

Scheuchzer's early career presented him with numerous examples of local natural history. He grew up in a family of physicians, and as a youth, he had often accompanied his father on botanizing expeditions. Both his father and grandfather had medical interests; and both had climbed the steep slopes of Mt. Pilatus above nearby Lucerne, a somewhat unusual venture at the time.[54] The younger Scheuchzer received a stipend from the city council of Zürich to attend medical school at Altdorf near Nuremberg, where he spent several years studying medicine and mathematics, the latter under Jakob Sturm, who at that time was revolutionizing the study of the mathematical and experimental sciences in the Nuremberg area.[55] While at Altdorf, Scheuchzer reportedly went on numerous excursions collecting both plants and fossils in the company of Johann Jakob Baier, who would later become a professor at Altdorf and write a regional mineralogy of Nuremberg.[56] Scheuchzer's progress towards the medical degree at Altdorf, though, was reportedly slow, since he spent so much time devoting himself to mathematics and the sciences, rather than medicine (though admittedly opportunities for actual dissection were few). On the urging of friends and relatives back in Zürich, he transferred to the University of Utrecht in the Netherlands, and there wrapped up his medical degree in four months flat.[57] To celebrate his acquisition of the doctorate, Scheuchzer then set off on a *peregrinatio academica* through the Holy Roman Empire before returning to Switzerland. During his trip, though, he had taken the opportunity to visit numerous other German universities and make contacts with medical professors there, to whom he promised specimens of Swiss plants once he got home.[58] Those he met in the course of his tour were to provide the core of the correspondence network he then established upon returning to Zürich, one which would eventually extend over numerous countries.[59]

During the course of his education and travels, Scheuchzer had picked up a strong interest in natural history, as well as ambitions for the future. Soon after his return to Zürich, he set about laying the foundations for

[54] Rudolf Steiger, *Johann Jakob Scheuchzer (1672–1733). 1. Werdezeit (bis 1699)* (Zürich: Leemann, 1927), 9–10.
[55] Steiger, 31.
[56] Steiger, 41, and Hans Fischer, *Johann Jakob Scheuchzer (2. August 1672–1623. Juni 1733), Naturforscher und Arzt* (Zürich: Leemann, 1973), 17; see Johann Jakob Baier, *Oryktographia Norica* (Nuremberg: Impensis Wolfgangi Michahellis, 1708), discussed in the previous chapter.
[57] Steiger, 67.   [58] Steiger, 74–78.
[59] Scheuchzer's *Stammbuch*, inscribed with the signatures, best wishes, and classical wisdom of the people he met during the years of his youthful peregrinations, provides one with a good sense of many of the kinds of individuals he would include in his correspondence network; see Zentralbibliothek Zürich, Ms. Z II 649.

what he ultimately hoped would be a natural history of Switzerland in its entirety. We may learn something about his goals and intentions by looking at his *Charta Invitatoria*, or "introductory document, with questions attached concerning the natural history of Switzerland," which he published in 1699 and sent off to friends and associates.[60] This pamphlet was an appeal to potential subscribers to, and participants in, his natural-historical venture. First laying out numerous reasons why, in his opinion, a "natural history of Switzerland" was sorely lacking, Scheuchzer then offered up a set of queries needing answering. Displaying considerable resemblance to those in Boyle's "General Heads for the History of a Countrey, Great or Small", Scheuchzer's queries were, however, substantially tailored to fit the Swiss situation.

From the very first words of Scheuchzer's *Charta Invitatoria* it is evident that the Royal Society's Baconian and internationalist goals, which Oldenburg had sought so hard to promulgate, formed a crucial part of Scheuchzer's conceptual apparatus. Scheuchzer opened his tract by praising "that illustrious Anglican society" (i.e. the Royal Society of London), "renowned throughout the globe," and "that great restorer of the sciences, Francis Bacon, Baron of Verulam." Under the auspices of Bacon's empiricist approach, maintained Scheuchzer, the English had succeeded in using natural history to establish a secure foundation for natural philosophy, "both among their countrymen and among foreigners, whether those living in England or those going abroad to the ultimate shores of the Eastern and Western Indies." Owing to the efforts of the English, Scheuchzer admiringly reported, "how much natural history has since come to light."[61]

In his admiration for the Royal Society of London, Scheuchzer was far from alone in the Holy Roman Empire; Sachs and many other members of the Academia Naturae Curiosorum wrote repeatedly, especially from the 1670s onwards, of their esteem (verging on envy) for the British organization and the royal patronage they thought had brought it funds.[62] But in his admiration for the figure of Francis Bacon himself, Scheuchzer was, in fact, expressing a sentiment not yet widely heard in the German-speaking world. Continental Europeans had, on the whole, remained relatively insulated from the fervor with which the English had taken up their countryman's philosophy; the French had remained devoted to the rationalist system-building of Descartes and his followers, while the Germans were soon to establish a fondness for the philosophy of their own compatriot Leibniz, as eventually institutionalized at the university level by Christian

---

[60] Johann Jakob Scheuchzer, *Charta Invitatoria, Quaestionibus, quae Historiam Helvetiae Naturalem concernunt, praefixa* (Zürich: n.n., 1699).

[61] Scheuchzer, *Charta*, 1.

[62] Barnett, 64–66 and 170–172. Alchemists also had a tendency to cite Bacon from time to time; Margaret Garber, personal communication.

Wolff.[63] In Switzerland, the citation of Bacon was not completely unprecedented; an earlier physician in Zürich, Johann Jakob Wagner, had indeed published a work which he labeled a "natural history of Switzerland," and which had professed an admiration of Bacon (while simultaneously displaying considerable credulity towards reports of flying dragons).[64] Scheuchzer thus had a number of exemplars to follow. Yet Bacon, for Scheuchzer, symbolized an approach to natural history that remained new, even almost a century after it had first been articulated. Offering him a series of ready-made templates for how to contribute to the new sciences – i.e. through the construction of his very own "natural history" – it also left him considerable scope for his own ambitions.

Scheuchzer thus took the Baconian program, and immediately began to modify it to fit his own circumstances. The English, he suggested, had specialized in learning about the natural objects of places far away, owing to their colonial empire. In contrast, he proposed that the Swiss turn their attention directly to their own non-colonial surrounds. "Because if those learned in physical truth pay attention to those things, which touch on the remote regions of the earth, by what greater right ought they to pay attention to those things in their native land; and we to those things in our native land, by accomplishing the earnest task of investigating those phenomena of nature, which she produces."[65] In Scheuchzer's view, the nature of Switzerland had been either ignored or denigrated by the rest of the known world. Quoting Aristotle in his defense ("nothing is sweeter than one's native land, however harsh and uncultivated it may seem"), Scheuchzer lept to the defense of his birthplace. "Our region is harsh, we agree, and indeed rough to the sight, but others should not believe it uncultivated, nor was it placed in an out-of-the-way corner of the world, where nature, tired from her labors, randomly and without any order cast those things which would either have been burdensome to other regions, or useless or indeed hardly necessary."[66] In Scheuchzer's defense of his native land, we can hear echoes of the defenses of *patria* invoked in the ongoing polemics around the "indigenous," as developed both in local floras and in regional mineralogies. In this defensive repudiation of negative natural stereotypes, we see Scheuchzer's strong interest in international opinion; he would write this book not just for the benefit

---

[63] For a rather narrow account of some British influences on German philosophy, see Gustav Zart, *Einfluss der englischen Philosophen seit Bacon auf die deutsche Philosophie des 18. Jahrhunderts* (Berlin: Dummler, 1881).

[64] Johann Jakob Wagner, *Historia naturalis Helvetiae curiosa* (Zürich: impensis Joh. Henrici Lindinneri, 1680), sig. *2v. A notice of the book's publication was printed in the Royal Society of London's *Philosophical Transactions*, 13 (1683): 268–271; the reviewer approvingly noted Wagner's references to "my Lord Bacon," but exhibited considerable skepticism about the dragons.

[65] Scheuchzer, *Charta*, 1.   [66] Scheuchzer, *Charta*, 2.

of the Swiss themselves, but in order to redeem Swiss natural history in the eyes of other lands.

Scheuchzer clearly intended his project to be novel in several ways. First would be the very scope of the book, covering "all sorts of particulars... those hidden in the land and in the sea, in the sky, in the air, in the earth, and most of all in that triple kingdom, vegetable, mineral, and animal."[67] Scheuchzer's 186 "*Quaestiones*," or "questions concerning the natural history of Switzerland" far exceeded in breadth those tackled by authors of local floras or regional mineralogies. They encompassed everything from the heavens to the earth. Scheuchzer asked his readers about alpine frost, about meteors, about dragons. As he did so, he showed his proclivity towards the new sciences by asking for precise measurements of all phenomena, whether of the pressure of the air, or the height of mountains. He asked what kinds of figures people had seen on snowflakes. He asked many questions about phenomena specific to the Swiss Alps, ones which were not known elsewhere. He asked questions about the Alps' inhabitants. He asked about rivers, lakes, and mountains, citing authors who had previously written on these topics, asking whether their information was still true. He suggested experiments to be performed upon, or with, Swiss mineral waters. He asked about the age of old men, and the fertility of women. He asked what kinds of "grains, herbs, fruits, and trees, whether of gardens or forests" the region held, and what use they had in economics, politics, medicine, and technology. He asked where they were to be found, and whether they differed in quality if they came from "dry fields, sandy fields, rocky fields, swampy fields, mountainous fields, or flat fields."[68] Finally, "since our native land is a milk-bearing region," he closed with a dozen questions about milk and cheese and their production.[69] In short, the breadth of Scheuchzer's interests, and his desire to integrate them into a single comprehensive framework, went far beyond what had become the conventional strategies of previous localist authors (and even, perhaps, those of many English Baconians).

Another novel feature of Scheuchzer's project in the Swiss context may be seen in the audience to which he aimed his queries. Far from addressing primarily his current or future medical colleagues, as local florists and regional mineralogists did, he explicitly called upon "curious men of the lowest order, fishers, shepherds of flocks, inhabitants of the Alps, farmers, diggers, root-cutters..." to contribute their observations, assuring them that he would not fail to cite them properly.[70] Remarking that he had been brought to this seemingly drastic step by the "great copiousness of the things to be observed," he explained that in order for him to undertake a work of this size, he would need to enlist the collective efforts of all, rather than merely relying on his own labors.[71] "For this work is certainly of such a

[67] Scheuchzer, *Charta*, 1.   [68] Scheuchzer, *Charta*, 11.   [69] Scheuchzer, *Charta*, 15.
[70] Scheuchzer, *Charta*, 2.   [71] Scheuchzer, *Charta*, 3.

great size that I could not presume to achieve it, unless I was helped by many people. For so great and of such a huge mass are the Alps of Helvetian nature.... "[72] He thus urged all "just and honest men of every order" to contribute to what he called "works of truth, works of our native land, works of nature."[73] While clearly drawing on Baconian ideals of collaboration, then, Scheuchzer gave them a distinctively Swiss twist. By making the Alps iconic, not only of Swiss nature but of the sheer size of his own endeavor, he was able symbolically to transcend the internal divisions within the Swiss Confederation. In this way, he aimed to persuade his readers that the "works of our native land" and the "works of nature" were indeed one and the same.

Scheuchzer's next step towards the realization of his project, a work he called his *Prolegomena to a Natural History of Switzerland*, approached the task from a somewhat different angle. Here Scheuchzer chose to begin his "busy inquiry" into "this most ample book of nature" by examining a different set of "books of nature," namely those volumes on natural history which had come his way so far. The bulk of the treatise thus consisted of a natural-historical bibliography.[74] But in the remainder of the treatise, Scheuchzer went down through the various categories of natural objects he planned to include, one by one, turning his expansive vision of natural objects in his previous pamphlet into a list of disciplines to be subsumed under natural history: meteorology, hydrography, mineralogy, botany, zoology, "galactology" (the study of milk), and anthropology (which he defined in this case as the study solely of the "inhabitants of Switzerland"), ending with the traditional catch-all category of "miranda" (things to be wondered at).[75] He stressed that his work would be both "narrative" (easy to understand), "inductive," and "applicative," with the latter facet instructing readers on how Scheuchzer's "theory of Swiss nature" could be put to uses medical and economic, technical and political.[76] Scheuchzer mitigated the overwhelming inclusiveness and ambition of his program, however, by making it seem *accessible* and within reach, in the process playfully mocking both the new sciences and contemporary fascinations with the exotic: "you won't need a telescope to see the things we're going to talk about in this book. Many things lie either near us or are close enough to be sought by our countrymen, and without great expense; we can see these as if in a mirror and can thus consider them more accurately."[77] In short, Scheuchzer promoted

[72] Scheuchzer, *Charta*, 2.   [73] Scheuchzer, *Charta*, 3.

[74] Johann Jakob Scheuchzer, *Historiae Helveticae naturalis prolegomena* (Zürich: typis Davidis Gessneri, 1700), 1. This work will be discussed further in the next chapter.

[75] Scheuchzer, *Historiae Helveticae naturalis prolegomena*, 20–25 (the last page of these appears, owing to an apparent printing error, as p. 23).

[76] Scheuchzer, *Historiae Helveticae naturalis prolegomena*, 2.

[77] Scheuchzer, *Historiae Helveticae naturalis prolegomena*, 27.

his project by uniting the broad ambition of an expanded Baconian "natural history" with the simultaneous promise of a goal tantalizingly within reach.

Scheuchzer had already begun a lengthier series of excursions into the mountains surrounding Zürich; in 1702, he was granted funds by the Zürich city council to continue these. In this way Scheuchzer gradually collected material for his natural history of Switzerland, until in 1706 he was able to begin its gradual publication. Rather than publishing in Latin, Scheuchzer chose to publish in German, thus aiming at a much broader audience than previous continental local natural-historical works had. Furthermore, he chose to publish his researches in what had been for naturalists, up till then, a most unusual format, that of the *weekly* publication, or magazine. Though the various scientific societies of the seventeenth century had indeed begun to publish regular journals (including the Academia Naturae Curiosorum, of which he was a member, and which published a yearly *Miscellanea* under various titles), they had not done so on nearly such a frequent basis.[78] Scheuchzer's model here seems rather to have been the weekly newspapers which, commencing even earlier in the seventeenth century, had by this point now begun to spread all over Europe. In this format, Scheuchzer began to tell "natural histories" (or "stories about nature"), each about a different natural phenomenon in Switzerland. Printing stories on several different topics each week, Scheuchzer ended up incorporating a vast array of materials into his natural histories. He regaled his readers, for example, with tales and legends about Switzerland's natural wonders, and with disquisitions on the medical problem of homesickness (to which he felt the Swiss to be especially prone), side-by-side with massive quantities of barometric measurements.[79] In these weekly installments for his natural-historical magazine, Scheuchzer drew on a much longer tradition of relating information about natural objects in the

---

[78] On the origins of scientific and scholarly journals, see David A. Kronick, *A History of Scientific and Technical Periodicals: The Origins and Development of the Scientific and Technical Press 1665–1790* (Metuchen, NJ: The Scarecrow Press, 1976); Sherman B. Barnes, "The Editing of Early Learned Journals," *Osiris* 1 (1936): 155–172; Otto Dann, "Vom *Journal des Scavants* zur wissenschaftlichen Zeitschrift," in *Gelehrte Bücher vom Humanismus bis zur Gegenwart*, ed. Bernhard Fabian and Paul Raabe (Wiesbaden: Harrassowitz, 1983), 63–80; and A. H. Laeven, *The 'Acta Eruditorum' under the Editorship of Otto Mencke (1644–1707): The History of an International Learned Journal Between 1682 and 1707*, trans. Lynne Richards (Amsterdam: APA-Holland University Press, 1990).

[79] Johann Jakob Scheuchzer, *Beschreibung der Natur-Geschichten des Schweizerlandes*, 57–60. Through his fascination with homesickness (or, in Latin, *nostalgia*) as a genuine and climatically-caused medical problem, Scheuchzer, like other Swiss authors before and after him, urged the medical understanding of place as central to a person's well-being. On later Scandinavian *nostalgia* see Lisbet Koerner, "Daedalus Hyperboreus: Baltic Natural History and Mineralogy in the Enlightenment," in William Clark, Jan Golinski, and Simon Schaffer, eds., *The Sciences in Enlightened Europe* (Chicago, IL: University of Chicago Press, 1999), 416.

form of stories or narratives.[80] Yet Scheuchzer made this too his own by adapting it to the abbreviated weekly format of his small magazine, which did not allow the presentation of longer "stories" in the same way large humanist volumes had. Scheuchzer's narratives were short and to the point, and told morals not so much about natural objects in themselves as rather about the *places* where these natural objects were to be found, united by their common location in Switzerland. Scheuchzer thus, in these writings, postponed his larger goal of providing any kind of complete natural history (what might be called "the" natural history of a place), but rather gave his readers a series of *individual* "natural histories," one after another, with the emphasis here clearly on the plural. In so doing, Scheuchzer departed yet again from the English model of natural-history writing typified by Plot, while opening up yet another alternative way to write the natural history of the territory.

Scheuchzer's weekly installments of "natural histories" came eventually to form the foundation for his *Helvetia historia naturalis, oder Natur-Historie des Schweitzerlandes*, an anthologized version of his journal which he ended up publishing in book form in multiple volumes.[81] Here Scheuchzer methodically rearranged his earlier randomly-ordered magazine installments into specific categories such as "hydrography," "meteorology," and "oryktography" (i.e. mineralogy), similar to the initial categories he had proposed many years before in his *Prolegomena*. They show Scheuchzer as having, at long last, fulfilled his initial goal of having arranged the natural history of Switzerland methodically, so as to include as many different varieties of natural phenomena as possible, from the study of Switzerland's mountains through the study of its air. The popularity of Scheuchzer's collected volumes on the natural history of Switzerland was such that they were ultimately reprinted many times, even after his death. In the process, Scheuchzer's works came to form a model for how a style of doing natural history originating within one specific national context could successfully be translated into another.[82]

---

[80] See for example Roger French, *Ancient Natural History* (London: Routledge, 1994). For discussions of the role of narrative in seventeenth- and especially eighteenth-century natural history, see Wolf Lepenies, *Das Ende der Naturgeschichte: Wandel kultureller Selbstverständlichkeiten in den Wissenschaften des 18. und 19. Jahrhunderts* (Munich: Hanser, 1976); Paul Lawrence Farber, "Buffon and Daubenton: Divergent Traditions within the *Histoire naturelle*," *Isis* 66 (1975): 63–74; and Phillip R. Sloan, "Natural History, 1670–1802," in *Companion to the History of Modern Science*, ed. R. C. Olby et al. (London: Routledge, 1980), 295–313.

[81] Johann Jakob Scheuchzer, *Beschreibung der Natur-Geschichten des Schweizerlandes; Helvetia historia naturalis; Natur-Geschichte des Schweizerlandes, samt seinen Reisen über die schweitzerische Gebürge* (Zürich: bey D. Gessner, 1746); *Natur-Historie des Schweitzerlandes* (Zürich: bey Heidegger und Comp., 1752).

[82] They also, of course, helped to create an enduring tradition of natural-historical interest within Switzerland itself, with many followers, for example Albrecht von Haller, who would soon expand the variety of natural-historical genres to include even poetry with his "Die Alpen," published in his *Versuch schweizerischer Gedichten* (Bern: Haller, 1732).

"THE POSSIBILITIES OF THE LAND"

In 1722, an obscure court physician published a short treatise on "the natural history of Saxony." Urban Gottfried Bucher offered the following reason why he had chosen to write a "natural," and not a more conventional civil history: "for how little one really knows about the changes among nations, and about their governance, and about their old boundaries. The detailed investigation of such things, in my opinion, brings little of use." In contrast, he argued, natural history was definitely of use, since it provided knowledge that he claimed was stable and reliable, knowledge that was based on a seemingly unchanging "nature," and that thus did not shift arbitrarily over time.[83] Unlike boundaries, which were human constructions, natural phenomena were enduring and offered power to the person who understood them. In short, Bucher found in the natural history of the territory a "useful" project, one that he thought might perhaps prove useful both to his prince and to his own career.

The case of the German territories affords yet another glimpse into how styles of natural-historical writing differed from place to place within Europe, and of how authors attempted to bridge these differences. During the early years of the eighteenth century, a number of German authors were to take inspiration from Scheuchzer's example, and to set out to write complete "natural histories" of their own territories. In the works they composed, frequently citing Scheuchzer but only very rarely Bacon, we can also see how these authors reformulated imported visions of natural history, just as Scheuchzer had, to fit their own needs. Though scattered across the Holy Roman Empire, these individuals shared several key goals. First, like Scheuchzer, they refused to limit themselves to the inherited disciplines of botany, mineralogy, and zoology, but rather reveled in the grand vision of natural history as going beyond the traditional three kingdoms. Second, they aimed their works at a "public" beyond the university, in most cases publishing in German so they could be understood by the locals themselves (including, of course, the princes and nobles whom they hoped to attract as potential patrons). They saw natural history as an enterprise not only conducted for the "public good," but also requiring public involvement. Here too, while following Scheuchzer, they made a radical departure from the local florists. Finally, however, they saw this enterprise as one naturally focused on the

---

[83] Urban Gottfried Bucher, *Sachßen-Landes Natur-Historie, oder Beschreibung der Natürlichen Beschaffenheit und Vermögenheit der zu Sachßen gehörigen Provinzen* (Pirna: druckts Georg Balthasar Ludewig, 1722), sig. A2r. The distinction between "civil" and "natural" history was, of course, a commonplace one in the early modern period; see for example Jean Bodin's presentation of it in his influential *Methodus ad facilem historiarum cognitionem* (Paris: apud Martinum Iuvenem, 1566), 9. What is especially interesting here, however, is the way in which Bucher explicitly attempts to justify his *preference* for natural over civil history, in the process investing both natural and civil history with new resonances.

geographical unit of the territory – as found not in England or Switzerland, though, but rather in the early modern Holy Roman Empire, where weak-to-nonexistent central authority did little to curb the ambitions of princes for their principalities. Here these authors departed from Scheuchzer and his English models. Eagerly appealing to German nobles for sponsorship, they attempted to frame natural history as a vehicle for princely power. This section shall investigate several of these imitators of Scheuchzer's, and discuss their varying successes and failures as they sought to define the "nature" of their territories.

It is crucial to understand how these authors viewed "natural history," in order to see why they set so much stock in it. As noted earlier, the way in which Bucher and others came to set "natural history" strongly against its "civil" counterpart is striking, and marks a clear break with the English tendency to incorporate the study of antiquities into natural history, as Plot and other English writers had done. Yet the ways in which the various authors defined "natural history" were complex, and sometimes contradictory. Take, for example, Bucher's own definition of the term in his book's very first sentence: "Natural history contains an account of that which grows or can be found in a territory (*Land*), and which serves the use and custom of its inhabitants."[84] Here, amidst Bucher's focus on perceived utility, it is not clear where nature begins or ends. Other authors granted their works a similarly broad scope. Michael Bernhard Valentini, though he pedantically defined natural history as the study of metals, plants, and animals (cf. the traditional three kingdoms) in the introduction to his *Prodromus historiae naturalis Hassiae* (Prodromus to a Natural History of Hessen, 1707), shed his scruples in the text itself and extended his work to all "natural bodies," from the "soil, mountains, and valleys" in Hessen, to its "air, winds, and waters," all the way up the great chain of being to its "brutes" or "quadrupeds."[85] Like Scheuchzer, then, whom all these authors cited repeatedly, they favored a definition of "natural history" expanded well beyond that to be found in the local flora and mineralogy. While Valentini, for example, on one occasion did refer in his book to a previous local flora, he made it clear that his own work was something on a completely different scale.[86] Johann Jockusch, author of a *Versuch zur Natur-Historie der Graffschafft Mannsfeld* (Attempt towards a Natural History of the County of Mansfeld, 1730), expressed frustration with previous authors who had left behind, he argued, nothing

[84] Bucher, 1.
[85] Michael Bernhard Valentini, *Prodromus historiae naturalis Hassiae* (Giessen: typis Henningi Mülleri, 1707), 1, 3, 7, 27.
[86] Valentini, 24. Likewise operating on this new, grander scale, though citing the mid-17th century English author Christopher Merrett's *Pinax rerum naturalium Britannicarum* rather than Scheuchzer, was a Polish Jesuit named Gabriel Rzaczynski in his *Historia naturalis curiosa regni Poloniae, magni ducatus Lituaniae, annexarumque provinciarum* (Sandomir: typis Collegii Soc. Jesu, 1721), sig. (3)v.

"sufficient and complete," leaving only "ruins" of works.[87] Rather than being disrespectful of the textual traditions they had inherited – indeed, most authors of "natural histories" liberally cited and quarried any and all works relating in any way to the study of their geographical area regardless of genre – they seem to have presented themselves as possessing a new kind of affinity for "nature," one which demanded that they account for nature's "completeness."

Scheuchzer's followers aligned themselves readily with the methods of the new sciences, and sought to incorporate them into "natural history," which they saw as equally avant-garde. Valentini attacked scholasticism in *"physica,"* and applauded a trend towards "ocular inspections, observations and experiments," which he saw as having greatly aided the growth of natural history.[88] Jockusch, meanwhile, was particularly interested in turning the concept of "experiment" to use in natural history. Repeatedly urging further "physical and especially economic observations and experiments," he provided examples for his readers of "experiments" which included introducing fish into a lake, and of botanical "experiments" in getting garden plants to grow, while encouraging his readers to send him accounts of their own garden trials.[89] Here as elsewhere, the natural history of the territory was presented as something new, as an enterprise capable of providing substantial benefits for anyone capable of appreciating its novelty and potential advantage.

Seeking patronage outside of the university context, the authors of these works dedicated them almost without exception to princes and nobles, and presented them as much-needed contributions to the moral and fiscal economy of the principality. Peter Wolfart, in his *Historiae naturalis Hassiae inferioris pars prima i. e. Der Natur-Geschichte des Nieder-Fürstenthums Hessen Erster Theil* (Natural History of Lower Hessen, First Part, 1719), dedicated his work directly to "Your Serene Highness CARL the first of this name, Count of Hessen."[90] Furthermore, he addressed the prince personally: "since your princely highness holds hidden treasures of the underground of this kind in no little esteem. . . . "[91] By publishing his book with text in both Latin and German simultaneously, Wolfart seems to have taken care to make his work accessible to multiple audiences, including not only scholars but those at court. While Wolfart addressed his prince directly, Jockusch, on the other hand, targeted instead various noble residents of the area, stressing

---

[87] Johann Jockusch, *Versuch zur Natur-Historie der Grafschafft Mannßfeld* (Eisleben: gedruckt bey Joh. Philipp Hüllmann, Consistorial- und Raths-Buchdrucker, 1730), 1, 4.

[88] Valentini, 1.   [89] Jockusch, 15–16 and 18.

[90] Peter Wolfart, *Historiae naturalis Hassiae inferioris Pars prima . . . i. e. Der Natur-Geschichte der Nieder-Furstenthums Hessen Erster Theil* (Kassel: gedruckt bey Heinrich Harmes, 1719), dedication.

[91] Wolfart, dedication. The first (and, as it turned out, only) volume he published was on the mineral kingdom.

their role in the administration of the territory: "you most noble and high men administer the both important and respected offices, entrusted to you by the highest authority, in the county of Mansfeld with the greatest fame."[92] He ascribed his hope of interesting local aristocrats in his project to Scheuchzer himself, who, he said, had "directed his plea not only at learned people, particularly physicians, who because of their profession have the most opportunity to attend to *Naturalia* better than others, but also at the nobility, who can supply the best accounts through the people on their own properties, especially administrators, foresters, fishers, shepherds, and old peasants."[93] Jockusch added that Bucher had, however, found that the latter groups did not always make the most "capable" informants, and that the proprietors of landed estates tended to be too busy to get involved; this had led Bucher to propose the clergy as an alternative audience.[94] Nonetheless, the appeal not only to princes, but also to noblemen and court bureaucrats seems to have been an attractive one for Scheuchzer's followers, many of them court physicians, as a source of patronage offering far greater resources than the university.

Jockusch thus suggested, following Scheuchzer, that the process involved in enlisting nobles and that involved in enlisting a wide variety of other occupations might be one and the same. In so doing, he clarified the potential role of the "public" that he and his fellow authors so commonly invoked when they claimed that their works were for the "public good."[95] Jockusch and his peers aimed not simply to instruct the "public" by popularizing the study of natural history in the vernacular, but rather to enlist this very public as a source of both information and support. In this way, patronage concerns blended with the need for collaborative work that Scheuchzer (and, of course, Francis Bacon before him) had seen as essential to the compilation of truly ambitious and thorough natural histories.

One of the most interesting individuals to answer Scheuchzer's call was Urban Gottfried Bucher, an obscure, medically-trained projector who in the early years of the 1720s worked his way back and forth between several German courts, attempting to sell his services as a "physicist" and naturalist to assorted princes. Very little is known about Bucher's life beyond his publications, but these are of considerable interest. In the supplement to Christian Gottlieb Jöcher's famous *Allgemeines Gelehrten-Lexicon* (General Index of Scholars), we learn nothing more than that Bucher "lived in the first half of the present [i.e. eighteenth] century,"[96] and wrote a book on the origins of the Danube (1720)[97] and a biographical account of the prominent early

---

[92] Jockusch, sig. [A1]r.   [93] Jockusch, 16.   [94] Jockusch, 17.

[95] See, for example, Jockusch, 16.

[96] Christian Gottlieb Jöcher, *Allgemeines Gelehrten-Lexicon* (Leipzig: J. F. Gleditsch, 1750).

[97] Urban Gottfried Bucher, *Der Ursprung der Donau in der Landgraffschafft Fürstenberg, samt des Landes Beschaffen- und Vermögenheit, untersuchet, und mit andern hierzu dienenden*

German cameralist and polymath discussed in the previous chapter, Johann Joachim Becher (1722).[98] What Jöcher and other bibliographers somehow missed was Bucher's third book: namely his *Sachßen-Landes Natur-Historie, oder Beschreibung der Natürlichen Beschaffenheit und Vermögenheit der zu Sachßen gehörigen Provinzen* (Natural History of Saxony, or Description of the Natural Qualities and Possibilities of the Provinces belonging to Saxony, 1722), which he then revised and expanded a year later. By examining his strange and curious career, we may learn much about the influences and motives behind the emergence of territorial "natural history" in Germany.[99]

Bucher's book on the origin of the Danube, in fact, displays many of the themes that were to emerge in his *Sachßen-Landes Natur-Historie*. Dedicated to the count of Fürstenberg, the book was framed as an attempt to provide literate proof that the river Danube originated in the count's own Black Forest domains. With its focus on "origins," the book may be seen to have strong links with the tradition of civil history, in the form of antiquarian and genealogical works which sought to sort out dynastic connections, and to trace the past boundaries of territories. Yet here Bucher sought the origins of a *natural* entity, the Danube river.[100] As we have seen, early modern naturalists themselves displayed yet another concern with "origins," in the

---

Physicalischen Anmerckungen auch einigen Oeconomischen Reflexionen (Nuremberg & Altdorf: bey Joh. Dan. Taubers sel. Erben, 1720).

[98] Urban Gottfried Bucher, *Das Muster eines Nützlich-Gelehrten in der Person Herrn Doctor Johann Joachim Bechers* (Nuremberg & Altdorf: bey Joh. Dan. Taubers seel. Erben, 1722).

[99] In addition to the works mentioned above, Bucher is also listed as respondent on the title pages of two medical dissertations, one dating from his early studies at Wittenberg (*Dissertationem medicam de catalepsi* (Wittenberg: ex Officina Goderitschiana, 1700)), another from his completed degree under Friedrich Hoffman at Halle (*Dissertatio inauguralis medica leges naturae in corporum productione et conservatione* (Halle: J. Gruner, 1707)). The authorship of a controversial medical-philosophical treatise on the soul (*Zweyer guten Freunde vertrauter Brief-Wechsel vom Wesen der Seele* (The Hague: bey Peter von der Aa, 1713)) has also been ascribed to him; for an attempt from the former German Democratic Republic at a Marxist interpretation of this work, as well as selected extracts from it, see G. Stiehler, ed., *Materialisten der Leibniz-Zeit* (Berlin: VEB Deutscher Verlag der Wissenschaften, 1966), 7–35, 177–199. More recently, Martin Mulsow has analyzed this treatise in the context of the complex interplay of ideas and influences in Wittenberg and Halle during the early German Enlightenment: see Mulsow, "Säkularisierung der Seelenlehre? Biblizismus und Materialismus in Urban Gottfried Buchers *Briefwechsel vom Wesen der Seelen* (1713)," in Sandra Pott and Lutz Danneberg, eds., *Säkularisierung in den Wissenschaften seit der frühen Neuzeit* (Berlin: Akademie-Verlag, 2002), 145–173. A reprint edition of the treatise in question, edited by Dr. Mulsow, will shortly be appearing in the series "Philosophische Clandestina der deutschen Aufklärung."

[100] For more on the search for natural as opposed to artificial boundaries, see Peter Sahlins, *Boundaries: The Making of France and Spain in the Pyrenees* (Berkeley: University of California Press, 1989); and Lucien Febvre, "*Frontière*: The Word and the Concept," in *A New Kind of History*, ed. Peter Burke, translated by K. Folca (London: Routledge & Kegan Paul, 1973), 208–218.

way in which they sought to name specific places and geographies as the origins of particular natural objects. By assigning the fount of one of Europe's most important rivers to the territory of this prince, Bucher was thus coming up with a formula for origins both natural and political.

Yet in many ways Bucher's book aspired not just to trace origins, but to provide a complete natural history of the territory in question. In Bucher's dedication, he labeled his book "a recension of natural things, occurring in the county of Fürstenberg."[101] Towards the book's end, he similarly described it as a "report" about "natural history." Here, Bucher defended his book and his investigation of "the often humble-appearing things within it," arguing that he had undertaken his research specifically for the benefit of the prince, who had wanted to learn about "the useful things hidden in the land for the benefit of his subjects."[102] In short, though his "report" had a specific political mission, Bucher made clear that he intended it to clarify the natural history of the prince's properties as a whole.

Bucher's broader interests are obvious from the contents of the book itself. Bucher wrote not only about the Danube's course, but about the atmospheric and meteorological features of the surrounding area, about the quality of its land, about farming and agricultural products, about its trees and wild game, its natural resources, mineral baths, and "materials" in general. And in so doing, Bucher earnestly took Scheuchzer as his model. On the very first page of the book, he started to critique previous geographers of the area, and of the German territories in general. Disparaging their qualitative methods, he approvingly cited the quantitative measurements of those modern *"physici"* who had begun to take instrumental readings of natural phenomena like atmospheric pressure and the height of mountains.[103] Here, Bucher named Scheuchzer as one of the best exponents of this new set of techniques, and giving an entire history of barometric measurement, commencing with the saga of Pascal's mountain ascent, he climaxed his story with Scheuchzer's dramatic measurements of Swiss peaks.[104] In these early sections of the book in particular, where he discusses questions of atmosphere and of the characteristics of mountains in the area, he is especially reliant on Scheuchzer for the form of "physical remarks and economic reflections."

One of the reasons Bucher placed particular emphasis on questions of measurement was, of course, that he wanted to establish the altitude of the area, in order to establish the prince's realm as the most likely source of the Danube. But he did not confine himself merely to proving this one point. He emphasized that his goal would be to look not so much at the "legal rights of the dynasty," but rather at "the situation of nature in the land."[105] Here Bucher seems almost to echo the wording of his statement, discussed

[101] Bucher, *Ursprung der Donau*, sig. [)(3]r.   [102] Bucher, *Ursprung der Donau*, 83–84.
[103] Bucher, *Ursprung der Donau*, 1.   [104] Bucher, *Ursprung der Donau*, 2–3.
[105] Bucher, *Ursprung der Donau*, 22.

above, that questions of boundaries ought to be separated from the study of nature. In other words, Bucher saw all aspects of nature as relevant to his "physical" and "economic" investigation of the land. Thus, discussing the "possibilities of the land," he attempted to characterize its "nature" broadly, arguing that because of the land's position high in the mountains, it was essentially best suited to raising cattle, and could not be faulted for its lack of other agricultural achievements.[106] He gave a detailed analysis of the current arrangements of men, meadows, forests, and livestock, examining their interactions.

But the land as it was was not the only thing that interested Bucher. He had visions for how the land could potentially be, if only "improved" with full attention to its economic "potential," and he shared these with the prince. To prevent the flooding of hay fields, for example, Bucher proposed a "projected canal."[107] Discussing the different kinds of crops that could and could not grow on the land, he argued for the strategic use of those crops best adapted to the area.[108] Bucher was concerned about the poor condition of local forests and, indeed, much of the Black Forest, owing to the chopping down of trees and the consumption of their wood for industry as well as for heat, as well as the neglect of native trees in favor of the cultivation of foreign ones, which didn't survive the winters. His proposed remedy for the fuel shortage, or "lack of wood," rested mainly in the protection of the woods, not only from their human population, but also from the cattle and wild game whose grazing harmed the trees.[109] Condemning the stop-gap solutions proposed by other economic advisors, he argued instead for the systematic reseeding of some areas.[110] In his attachment to these and other "projects," he may be seen as drawing on the cameralist traditions discussed in the previous chapter.

---

[106] Bucher, *Ursprung der Donau*, 36.

[107] Bucher, *Ursprung der Donau*, 40. On projects of domestic "improvement" (*Verbesserung*) in the eighteenth-century Holy Roman Empire, see Ulrich Troitzsch, *Ansätze technologischen Denkens bei den Kameralisten des 17. und 18. Jahrhunderts* (Berlin: Duncker & Humblot, 1966); Günter Bayerl and Torsten Meyer, "Glückseligkeit, Industrie und Natur-Wachstumsdenken im 18. Jahrhundert," in *Umweltgeschichte: Methoden, Themen, Potentiale*, eds. Günter Bayerl, Norman Fuchsloch, and Torsten Meyer (Münster: Waxmann, 1994), 135–158; and Martina Kaup, "Die Urbarmachung des Oderbruchs. Umwelthistorische Annäherung an ein bekanntes Thema," in the same volume, 111–133. For some non-Continental examples of the rhetoric of "improvement," see Richard Drayton, *Nature's Government: Science, Imperial Britain, and the 'Improvement' of the World* (New Haven, CT: Yale University Press, 2000); Simon Schaffer, "The Earth's Fertility as a Social Fact in Early Modern Britain," in *Nature and Society in Historical Context*, eds. Mikuláš Teich, Roy Porter, and Bo Gustafsson (Cambridge: Cambridge University Press, 1997), 124–147; and Karen Oslund, "'Nature in League with Man': Conceptualising and Transforming the Natural World in Eighteenth-Century Scandinavia," *Environment and History*, 10 (2004), 305–325.

[108] Bucher, *Ursprung der Donau*, 46–47.   [109] Bucher, *Ursprung der Donau*, 52.

[110] Bucher, *Ursprung der Donau*, 57.

Yet Bucher's cameralism was not solely administrative in focus, but displayed clear links with the kinds of natural-historical description that had previously been developed in Central Europe, in particular in the regional mineralogy. These may be seen most clearly in his discussion of the various mineral resources of Fürstenberg. Bucher maintained that Fürstenberg's mountainous terrain should be seen not as a drawback, but rather as an opportunity for the exploration of mineral wealth. He was confident that the area, in fact, possessed a full complement of natural resources, the importance of which regional mineralogies had begun to stress. Bucher asserted that the economically-minded observer must not ignore those objects "otherwise held as being limited," and must not "pass by those truly useful," lest he miss the true value of a territory.[111] He enthusiastically listed the numerous minerals to be found in the territory, emphasizing, for example, that limestone was so abundant that the castle had been built with it, while gypsum was to be found in numerous places.[112] He pointed out that there were great sulfur waters that were not yet being used by anyone.[113] Likewise he noted that the region had an abundance of different kinds of iron, scattered in different places. Although this local iron had previously not been used as much as it could have, because it was felt to be too brittle, Bucher felt that all that was needed would be to find a good "treatment" for it, or procedure that would enable the smelting of better-quality iron; technology could be set to work to make previously useless objects useful.[114] Finally, he concluded by arguing that it was not yet known whether other metals were in existence in the territory, and he urged the local miners to set to work, in order to figure out what metals *could* be found.[115]

In short, Bucher used the opportunity provided by this book to supply a comprehensive account of the natural contents of the territory of Fürstenberg. The way in which he provided this account may be seen to have been shaped by his own goals – ones which he sought to impress upon the prince – of finding "physical" and "economic" uses for it. Though he discussed the question of the origin of the Danube at great length, his main point in this discussion was to stress the value of the Danube as a *natural* phenomenon, one which could be treated by methods differently than those that geographers had used before. He constantly insisted on the importance of an accurate knowledge of the contents of local regions, foreshadowing his work on the natural history of Saxony. This knowledge could be put to use most directly in developing "projects" for territories, but it could also contribute to the larger intellectual world in which *physici* shared knowledge and ideas. This is clear, for example, in a passage where he discusses a fossil, noting current debates about the origins of fossils, and in particular English

---

[111] Bucher, *Ursprung der Donau*, 61.   [112] Bucher, *Ursprung der Donau*, 61–62.
[113] Bucher, *Ursprung der Donau*, 65.   [114] Bucher, *Ursprung der Donau*, 69–75.
[115] Bucher, *Ursprung der Donau*, 73.

mineralogist Martin Lister's interpretations of them as "jokes of nature," the result of plastic, creative forces within the earth. Using the fossil in question as a resource with which to argue against the famous English scholar, Bucher insisted instead that this very piece of evidence found in the county of Fürstenberg provided a proof – and a *local* proof at that ("a certain native proof") – in favor of the opposite hypothesis, namely that of the organic origin of fossils.[116] Natural objects within a prince's or nobleman's territories could thus serve not only practical and politico-symbolic purposes, but intellectual ones as well.

Bucher's next project, *Das Muster eines Nützlich-Gelehrten in der Person Herrn Doctor Johann Joachim Bechers* (The Pattern of a Usefully Learned Person in the Figure of Doctor Johann Joachim Becher, 1722), offered a biographical explication of the life and works of Johann Joachim Becher, the great German cameralist and projector of the mid-seventeenth century.[117] This work makes clear, in more ways than one, Bucher's cameralist involvements. In this book, Bucher basically rewrote Becher's life to suit his own interests and projects, including those in natural history. Crucial to Bucher's presentation of Becher was his labeling of him as a "usefully learned man." In this phrase, which occurs not only in the book's subtitle but repeatedly throughout, Bucher sought to establish a new model for the scholar, one focused not solely on the university life, but on making a contribution to the "common good."[118] In his first chapter, Bucher argued that Becher counted, in fact, as a "learned man," even though Becher practiced at assorted German courts, and was not affiliated with universities (beyond the medical degree he had attained early in his career). Bucher maintained that the inability to fit Becher into limited academic categories was precisely why he had earned his label of being a "*usefully* learned" person. Becher, Bucher argued, had displayed in his career "a particular affection towards the common German good."[119] He stressed those aspects of Becher's *curriculum vitae* which presented him as investigating the natural products of Austria, and arranging shipping routes for getting these to Holland.[120] Emphasizing Becher's fondness for "observations" and "experiments," he outlined the latter's numerous projects (including not only his perpetual motion machine, but also saw- and water-mills, textile machinery, ovens, and weapons).[121] Basically, what Bucher appears to have been most impressed by was Becher's sense of

---

[116] Bucher, *Ursprung der Donau*, 81.

[117] Urban Gottfried Bucher, *Das Muster eines Nützlich-Gelehrten*. For a more recent biography of Becher, see Pamela H. Smith, *The Business of Alchemy: Science and Culture in the Holy Roman Empire* (Princeton, NJ: Princeton University Press, 1994).

[118] See Walther Merk, *Der Gedanke des gemeinen Besten in der deutschen Staats- und Rechtsentwicklung* (Darmstadt: Wissenschaftliche Buchgesellschaft, 1968, c1934).

[119] Bucher, *Das Muster eines Nützlich-Gelehrten*, 25.

[120] Bucher, *Das Muster eines Nützlich-Gelehrten*, 21–22.

[121] Bucher, *Das Muster eines Nützlich-Gelehrten*, 51.

nature as constituting a "factory in the earth," a way of providing all of the goods which humans could possibly need, should they only apply their own industry to the extraction and proper manufacture of these goods.[122]

In Bucher's *Sachßen-Landes Natur-Historie* (1722), we see, in effect, the culmination of many of the themes of his two previous works. In this book, which he explicitly framed as a "natural history," Bucher set out to tell a set of stories about the "natural possibilities" of Saxony.[123] And, as in his previous works, he declared that he would pattern his own natural history after Scheuchzer's: "I will handle this history of nature not in an orderly series, but because our knowledge comes only piecemeal, I will recount it in that manner and by imitating my most worthy patron, Dr. Scheuchzer in Switzerland."[124] As the phrase "piecemeal" suggests, Bucher, like Scheuchzer and his own German compatriots, reported that he planned to deliver his natural history in multiple installments; however, for only one of these subsequent installments does the financing seem to have been forthcoming.[125] Here too, though, we can see some of the ways in which Bucher's style of writing natural history was *not* merely patterned directly on Scheuchzer's, but was rather very much adapted to the German situation of the princely court.

This much is evident from his "first report" which was "about the pleasantness of the region around Dresden and about a camphor tree growing there." He began his report with an immediate evocation of the princely status of Dresden as a city: "if one could find a pleasant region anywhere, one would find it around the royal Polish and princely Saxon capital city of Dresden."[126] And indeed, it is the royal presence in the city that is at the center of this report. Bucher commenced his report with a description of the city's position in a valley, surrounded by higher hills; and he went on to discuss which of these spots gave the "best views" of Dresden as the sight of princely power, even going so far as to compare one of them to an "observatory from which one can see the entire valley and even as far as Meißen."[127] Scattered throughout the slopes of the hills he mentioned the various royal pleasure villas. He concluded his description of the landscape by remarking that all of these different spots from which one could see the city only served as means of focusing attention back on the royal presence within that city: "surrounded by these pleasure-works of nature and art, like jewels, the royal residence of Dresden, which as the most inner and centrally-placed precious

[122] Bucher, *Das Muster eines Nützlich-Gelehrten*, 71.

[123] Bucher, *Sachßen-Landes Natur-Historie*, title page; the subtitle runs "*oder Beschreibung der Natürlichen Beschaffenheit und Vermögenheit der zu Sachßen gehörigen Provinzen.*"

[124] Bucher, *Sachßen-Landes Natur-Historie*, 8.

[125] Essentially an expanded version of the first edition, this was published as *Sachsen-Landes Natur-Historie, In Welcher Dieses Landes, und der darzu gehörigen Provintzen Natürliche Beschaffenheit, Vermögenheit und Begebenheiten, in unterschiedenen Erzehlungen vorgestellet werden. Erste Erzehlung* (Dresden: bey Joh. Christoph Krausen, 1723).

[126] Bucher, *Sachßen-Landes Natur-Historie* (1722 ed.), 9.   [127] Bucher, 10.

stone in this jewel, can be viewed from the palaces, just like one can, from the center, view many of these palaces resembling private houses as a most pleasant periphery."[128]

Bucher thus framed his discussion of Saxony's fertility, or "completeness," as a reflection of princely glory. Whereas much of the "natural history" he presented was indeed, for one influenced by Scheuchzer, fairly standard fare – he surveyed the situation of metals and rocks in the area, as well as agriculture, wildlife, wine, and even fish – nonetheless it was placed within a sharply different context than that of bourgeois Zürich. Bucher shaped his descriptions to appeal to princely taste, for example, describing even the *sounds* of the place as glorious: "between these cliffs there is the most beautiful echo, and in the early year the most pleasant nightingale's song, but before all things this valley of the Elbe recommends itself with the most temperate air."[129] Bucher cleverly used his discussion of local atmospheric conditions to segue to his final section. Here he praised the Dresden climate still further, by discussing the case of an imported camphor tree which had somehow succeeded in flourishing there: "And this is the reason for the uncommonly well-constructed gardens around here, and for the progress of the otherwise most delicate garden plants with which our Dresden will soon rival Italy. I won't even try to describe the royal *Orangerie* and its excellence, but for the moment will remain in the garden of the royal Polish and princely Saxon administrator, Herr Heig, who near the *Orangerie* among other exotic plants has displayed an East Indian camphor tree, which the gardener Carl Augadt got to bloom."[130] Bucher thus closed his "first report" with yet another invocation of court power, this time seen as having command not only over territorial productions but over the external world as well. Even "exotic" species could, in the natural history of the territory, be used to serve the overall goal of promoting princely power.

How, then, to reconcile Bucher's lavish praise for the Dresden court with his declaration, mentioned earlier, that "natural history" could and should be freed from the uncertainties of civil history? For in his introduction, Bucher did seek to ground his "natural history" of Saxony on the idea that writing about the "nature" of an area might provide a far more secure foundation than any kind of writing about its human history. Here he had indeed contrasted natural and civil history, presenting the former as stable, with the latter fickle and subject to alteration: "[natural history] is that much more believable, the more unchanging nature is in her arrangements, good fortune of which civil history can not boast."[131] Bucher complained about the vagaries of human history, citing as examples the ancient German tribes, with their frequent wanderings and migrations, which had made it difficult for

[128] Bucher, 12.   [129] Bucher, 14.   [130] Bucher, 14.
[131] Bucher, *Sachßen-Landes Natur-Historie*, 3.

subsequent writers to locate any pure "nation" or "tribe."[132] Even though writers of constitutions, he argued, always sought a pure genealogy on which one could ground laws, the study of geography dissolved any notions of such purity.[133] In his rejection of the antiquarian approach to natural history fostered by Plot and other English writers, Bucher seems to have attempted to forge a more locally acceptable style of doing natural history, one more in keeping with the tradition of the local flora, for example, which *had*, unlike previous topographical works, focused solely on the natural to the exclusion of the civil; and one which was simultaneously conducive to his efforts to seek patronage at court as someone offering something *different* than traditional antiquarian researches. At any rate, the majority of these German projects must be termed failures, in that no subsequent installments appeared, and their authors vanished into mists of obscurity. Yet their example shows us once again the roots of contemporary interest in the compilation of territorial natural histories – and the difficulties in achieving communication across boundaries of place and conventions of style.

During the seventeenth and eighteenth centuries, then, different visions of "natural history," and in particular the natural history of the territory, came to be debated in a variety of places and contexts. Ultimately, as the chapter shows, regionally different traditions of natural-historical writing conditioned the acceptance, and altered the subsequent form and message, of the territorial natural history. Despite Oldenburg's best efforts, the English model, typified by Robert Plot's county natural history of Oxfordshire, was not to be easily reproduced elsewhere in Europe. As Oldenburg's correspondents repeatedly explained to him, their local situations and the expectations they themselves held about the proper practice of local natural history created gaps too great to bridge. Only much later did Johann Jakob Scheuchzer came to craft a form of Baconianism suitable for continental Europe, and even then confined mainly to the Swiss Confederation. Only by following Scheuchzer's example – long after Oldenburg's death – that German authors finally tackle the natural history of the territory, in the process, however, once again subtly transforming its goals and aims. Amidst the diversity of European contexts, the very place-based nature of natural-historical information made its transmission difficult. Only through the construction of new and alternative ways of communicating knowledge could this knowledge be incorporated into the new "universal" sciences of nature.

---

[132] Bucher, *Sachßen-Landes Natur-Historie*, 4.

[133] As he complained, "it is difficult to determine the old boundaries of the land, since at the time one really didn't have any commissioners of boundaries yet." Bucher, *Sachßen-Landes Natur-Historie*, 7.

5

Problems of Local Knowledge

On February 11, 1705, Johann Jakob Scheuchzer told a "natural history": a story about some red beech trees in the neighborhood of Zürich. The story he told was one about nature and the native, and about the relationship between different forms of knowledge about these. The story began by recounting the normal course of nature. It is well known, Scheuchzer remarked, that most trees' leaves turn red in the fall; but "in the neighborhood of Buch... there are among the other beech trees, oak trees, and other trees of the forest, three beech trees, which diverge from the common sort known in all of Europe...."[1] The trees were unusual in the following respect: "that they put on their colorful dress at the beginning of summer, and around the holy feast of Pfingsten they display a marvelously beautiful red, so that farmers living in the region for two hours [journey] around often collect themselves around the trees, so that they can break off leaves and branches from these blood-red trees, and so that they can wear these on their hats back home."[2] Even more remarkable than this unseasonably early changing of leaves' color was the fact that seedlings from the same trees would not repeat the feat elsewhere. Scheuchzer drew a moral from this story: that of the enduring relationship between creatures and their native land. "What's astonishing here is that which the neighbors of these trees report, that when they're transplanted to other places they don't grow, as if they were only interested in living in the earth in which they were born, or were refusing all other nourishment. So here it's not the case that *Patria est, ubicunque bene* [which might be translated, perhaps, as "one's fatherland is wherever things go well"], because these trees don't prosper anywhere except in their fatherland (*Vatter-Land*)."[3]

In this account we see Scheuchzer, the learned physician from Zürich, taking an example of local peasant custom, and making it into a story. What is especially interesting here, however, is how he then proceeds to break down his story. Scheuchzer's wonder at this natural phenomenon, and at its

---

[1] Johann Jakob Scheuchzer, *Beschreibung der Natur-Geschichten des Schweizerlandes* (Zürich: in Verlegung des Authoris, 1706–1708), 2. This was Scheuchzer's very first contribution to his weekly natural-historical periodical.

[2] Scheuchzer, 2.   [3] Scheuchzer, 2.

inability to be transported from one place to another, but rather its strict and literal rootedness in time and place, is matched by his wonder at the kinds of stories that the peasants themselves told to explain this phenomenon.

It should however cause great astonishment, what the inhabitants of the area have reasoned out about such an uncommon occurrence. They maintain that a long time ago four or five brothers were murdered at this place, and that according to the proper judgment of God five beech trees grew there with drops of blood on them as a monument to such a horrible deed. In this the farmers seem to be philosophers... [4]

Scheuchzer was skeptical about this story.

Here, however, both we and they are lacking sufficient proof to make this story a believable one. Nobody knows anything about the time when this murder story is supposed to have happened, or about the actors themselves, or about other circumstances necessary for establishing the truth of a history. But this is certain, that the farmers alive today could not have made up such a fable, but probably inherited it from their ancestors as a tradition. [5]

Instead, to explain this phenomenon, Scheuchzer turned to a comparison, to an example he had read about in an English county natural history: "I think that these blood-red beech trees should be compared with the similarly colored leaves of a birch tree which is to be found in a forest near Ranton Abbey in the county of Staffordshire, which likewise in the spring appears red, as if fresh blood had been poured on it, so that it also has been named the blood-red birch tree, according to the testimony of Robert Plot's *Natural History of Staffordshire*." [6] Scheuchzer, however, was not content merely to cite the testimony of another local naturalist, but developed his own physiological/circulatory explanation for the presence of the bright red in the trees' leaves: "Both these birch and beech trees probably have their wood vessels so closely pressed together, that only the subtle parts of the sap can ascend through the nutritive vessels, for which reason the vessels of the leaves cannot be extended so far that they will receive a green color like those of other trees." [7] With this rationalized, scientific explanation, which drew on his training both as a physician and as a follower of the new sciences, Scheuchzer closed his story. He had first considered the testimony of peasants and indeed paid great attention to it; and then he had rejected it. He had taken the occasion to spin a tale about plants and their habitats, and to consider the relationship between organisms and place. Yet in the process, he had also brought up a much larger issue: that of the relationship between the inhabitants of a place, and their knowledge of it.

Historians of science have just begun to examine the ways in which European naturalists, as they traveled out from Europe to places all around the world, framed their encounters not only with indigenous plants and animals,

---

[4] Scheuchzer, 2.   [5] Scheuchzer, 2.   [6] Scheuchzer, 2.   [7] Scheuchzer, 2.

but also with the indigenous peoples they met. Scholars have thus, for the first time, begun to take a serious look not only at what European travelers thought of local nature, but at the ways in which they used, acknowledged – and failed to acknowledge – the natural knowledge of the peoples whom they encountered. The focus in this chapter, as in the book as a whole, is not to study European interactions with the indigenous knowledges of Asia, Africa, and the Americas; this highly promising area of study, on which much more needs to be done, has recently been treated in a number of other works.[8] Instead, this chapter will explore, through several important case studies, how European naturalists framed their relationships with their own indigenes: that is to say, with the various kinds of collaborators and informants with whose assistance they wrote their local natural histories. The chapter will thus look at the numerous learned and popular (sub)cultures of Europe almost as if they were "indigenous" in their own right – and as we have seen, in many ways this is exactly how European naturalists viewed them.

Problems of local knowledge were highly significant for the writing of local natural history, since inevitably, much of naturalists' knowledge was gained from local people in one way or another, whether through generations of countryfolk who had found that a particular herb was useful for a particular ailment, or through farmers or stone-cutters who happened upon interesting fossils in the course of their daily work. Local natural histories drew explicitly on knowledge gained from inhabitants of specific areas. Yet they displayed a curiously ambivalent attitude towards this knowledge. For in fact, many competing kinds of knowledge about place and locality existed during the early modern period. Reference has already been made, in preceding chapters, to some of the various different kinds of folk or practical knowledges: for example, those possessed by wise women, apothecaries, gardeners, stone-cutters, miners, and those often called the "people" or, more pejoratively, the "vulgar."[9] The knowledge of a place possessed by its inhabitants often might differ greatly from that of outsiders. Learned knowledge, furthermore, was as heterogeneous as folk varieties, despite the

---

[8] See for example Richard Grove, *Green Imperialism: Science, Colonial Expansion and the Emergence of Global Environmentalism, 1660–1880* (Cambridge: Cambridge University Press, 1994), especially 73–94; Londa Schiebinger, *Plants and Empire: Colonial Bioprospecting in the Atlantic World* (Cambridge, MA: Harvard University Press, 2004); and the essays in Londa Schiebinger and Claudia Swan, eds., *Colonial Botany: Science, Commerce, and Politics in the Early Modern World* (Philadelphia: University of Pennsylvania Press, 2005).

[9] On lay medical knowledge see for example Paul Diepgen, *Deutsche Volksmedizin: Wissenschaft, Heilkunde und Kultur* (Stuttgart: Enke, 1935); Sergius Golowin, *Die Weisen Frauen: Die Hexen und ihr Heilwissen* (Basel: Sphinx, 1982); and Christian Probst, *Fahrende Heiler und Heilmittelhändler. Medizin vom Marktplatz und Landstraße* (Rosenheim: Förg, 1992). On the "Hausväterliteratur," see Julius Hoffmann, *Die Hausväterliteratur und die Predigten über den christlichen Hausstand* (Weinheim: Beltz, 1959).

relative unanimity among Enlightenment scholars in denouncing the latter variously as "superstitious" or in other ways suspect.[10] The new sciences had, for example, supplied a plethora of new ways of investigating and measuring place, such as mathematical geography and surveying.[11] Meanwhile, Aristotelian and humanist traditions, descended from classical models but following different trajectories, developed various verbal approaches to describe the history and topography of lands.[12] And as we have seen, natural history itself took various different paths, as authors developed their own intricate languages of description, often scarcely comprehensible to the folk who lived in the areas being described.

In short, natural history in the early eighteenth century was at a crossroads. With the arrival of various different ways of writing natural history – including the local flora, regional mineralogy, and local natural history among others – there were numerous ways in which one could frame the local natural world. Furthermore, there were equally numerous ways to decide how to view what we might term the "locals" who inhabited that world. This chapter will seek to probe the relations between the forms of knowledge seen in local natural history, and these various other knowledges, both folk and elite, upon which it drew and with which it intersected and often clashed. As naturalists strove to develop their own new languages for describing the natural world, they encountered previous and other languages about the land. By exploring the ways in which naturalists chose to acknowledge the works of others, we can see who was chosen to be included in this process, and who was not.

This chapter will explore these problems of "local knowledge" by examining several bibliographies of natural history that were published during the first half of the eighteenth century, namely those of Johann Jakob Scheuchzer

---

[10] On learned attitudes towards popular beliefs about natural phenomena and natural wonders, see Natalie Zemon Davis, "Proverbial Wisdom and Popular Errors," in *Society and Culture in Early Modern France* (Stanford, CA: Stanford University Press, 1975), 227–267; and Lorraine Daston and Katharine Park, *Wonders and the Order of Nature, 1150–1750* (New York: Zone Books, 1998). Also see Giuseppe Cocchiara, *The History of Folklore in Europe* (Philadelphia: Institute for the Study of Human Issues, 1952), for discussions of the ambivalence of the learned who eagerly sought out information about popular beliefs, while simultaneously subjecting them to learned scrutiny.

[11] See for example David Buisseret, ed., *Monarchs, Ministers and Maps: The Emergence of Cartography as a Tool of Government in Early Modern Europe* (Chicago, IL: University of Chicago Press, 1992); Svetlana Alpers, *The Art of Describing: Dutch Art in the Seventeenth Century* (Chicago, IL: University of Chicago Press, 1983; and David Buisseret, ed., *Rural Images: Estate Maps in the Old and New Worlds* (Chicago, IL: University of Chicago Press, 1996).

[12] The literature here is vast, so I will cite only Gerald Strauss, *Sixteenth-Century Germany: Its Topography and Topographers* (Madison: University of Wisconsin Press, 1959), on topography, and Peter Hanns Reill, *The German Enlightenment and the Rise of Historicism* (Berkeley: University of California Press, 1975), on history.

(whom we have already encountered in the preceding chapter) and of his famous successor Linnaeus, the originator of modern botanical nomenclature. During the sixteenth and seventeenth centuries, though naturalists had traditionally included lists of works consulted and had sometimes discussed the relative merits of these, perhaps including brief historical reflections on the progress of natural history as well, they had rarely seen fit to publish these lists as self-standing and entirely bibliographical volumes.[13] This, however, is what Johann Jakob Scheuchzer chose to do in the several versions of his *Bibliotheca scriptorum historiae naturali omnium terrae regionum inservientium*, or "library of writers of natural history serving all the regions of the earth" (1716). Linnaeus soon followed with a work aimed at providing a similar overview of the field for botanists, his *Bibliotheca botanica* or "botanical library" (1736). Various others were to compose similar works over the course of the eighteenth century, to help themselves and others master a growing literature. Though these works were not themselves "histories" in the modern sense, that is to say the fully-fledged histories of the various disciplines of natural history that were to emerge in the late eighteenth and nineteenth centuries,[14] they nonetheless offered their readers insight into their authors' respective visions of the past, present, and future of natural history. The chapter will explicitly contrast Scheuchzer's and Linnaeus's bibliographies, as representing two very different styles of approaching local knowledge. In these works' varying valuations of local informants, the chapter will argue, may be seen a series of struggles about the ultimate role that different local knowledges would play in shaping what would be seen as "scientific" knowledge of the natural world.

"THE INDIES IN SWITZERLAND"

In 1700, Johann Jakob Scheuchzer published a preliminary bibliographical survey of writers on the natural history of Europe, the Americas, Africa, and Asia, in a pamphlet titled "prolegomena to the natural history of

---

[13] On historical musings in early modern works on natural history and astronomy, see Peter Dilg, "Vom Ansehen der Arzneikunst: Historische Reflexionen in Kräuterbüchern des 16. Jahrhunderts," *Sudhoffs Archiv* 62, 1 (1978): 64–79; Joseph M. Levine, "Natural History and the History of the Scientific Revolution," *Clio* 13 (1983): 57–73; Nicholas Jardine, *The Birth of History and Philosophy of Science: Kepler's 'A Defense of Tycho Against Ursus'* (Cambridge: Cambridge University Press, 1984).

[14] For an introduction to the context of the emergence of modern histories of the individual sciences, see Rachel Laudan, "Histories of the Sciences and Their Uses: A Review to 1913," *History of Science* 31 (1993): 1–34; and Loren Graham, Wolf Lepenies, and Peter Weingart, eds., *Functions and Uses of Disciplinary Histories* (Dordrecht: Reidel, 1983).

Switzerland" (*Historiae Helveticae naturalis prolegomena*).[15] In 1716, he revised this and published it in much-expanded book form, as a "library of writers of natural history serving all the regions of the earth" (*Bibliotheca Scriptorum Historiae Naturali omnium Terrae Regionum inservientium*).[16] In the story of this book, whose pages gave him ample room to comment on the contributions and faults of numerous authors, we can see how Scheuchzer constructed his own vision of the growth of the natural-historical enterprise. In particular, we can see how Scheuchzer, in keeping with his own advocacy of the particular genre of the natural history of the territory, formulated an extremely inclusive history of natural history, one singularly open to the acknowledgement of many different people and places. This section will examine Scheuchzer's bibliography, however, not so much for the light it sheds on his own career, as rather for the ways in which it articulates an implicit vision of the status of local natural history and of local informants within it. In the *Bibliotheca*, Scheuchzer expressed a strong commitment to the project of local natural history and to the local informants who made it possible, a view that he shared with others at the time, but that he expressed with particular vigor. While noting the limits and limitations of Scheuchzer's inclusiveness, then, I will attempt to draw out the implications of the ways in which he claimed local study as the center of the enterprise of natural history, and local knowledge as the heart of knowledge-making itself.

In Scheuchzer's *Bibliotheca*, in fact, while some herbals and taxonomic works were listed, works of local natural history – local floras and mineralogies, and natural histories of individual territories – predominated by far. Scheuchzer at one point referred to the bibliography as one of "special natural history," that is to say the natural history of particular places (as opposed to "general natural history," under which heading systematic works usually fell).[17] From the very first version of the bibliography in his *Prolegomena*, Scheuchzer chose to organize it by region, with separate sections covering each geographical entity, from "Helvetia" (Switzerland) and "Germania" (presumably the Holy Roman Empire; unfortunately, he did not explain his use of the term) through "Asia," "Africa," and "America" where the categories he used came to encompass larger and larger areas. What is striking in Scheuchzer's catalogue of books is that, just as we have seen in contemporary

[15] Johann Jakob Scheuchzer, *Historiae Helveticae naturalis prolegomena* (Zürich: typis Davidis Gessneri, 1700).

[16] Johann Jakob Scheuchzer, *Bibliotheca Scriptorum Historiae Naturali omnium Terrae Regionum inservientium* (Zürich: typis Henrici Bodmeri, 1716). The book was reprinted under the same title in 1751. The Handschriftensammlung in the Zentralbibliothek Zürich also contains a set of post-1716 revisions to the *Bibliotheca*, which Scheuchzer seems to have added by hand onto pages of the work glued into a larger blank volume; see Zentralbibliothek Zürich, Ms. Z VIII 18.

[17] Scheuchzer, *Bibliotheca*, 1. Here he refers to the volume as a "library of writers on special natural history."

catalogues of plants and other natural objects, he felt the need to segregate these books from each other geographically, separating the "indigenous" from the "exotic," but on a far more refined and complex scale. Not only did Scheuchzer devote the bulk of the book to localist works, then, but he also stressed their localism as precisely the feature that made them valuable to him.

Scheuchzer introduced his *Bibliotheca* as a work which had indeed originated with his specific interest in the natural history of Switzerland, as the title of his *Historiae Helveticae naturalis prolegomena* suggests, but which in the process had ended up probing the furthest corners of the world: "... this book indeed does not remain within the limits of our tiny little mountainous country, but because my curiosity is so insatiable, it extends to Ultima Thule, to the Garamantas and to the Indies...."[18] Scheuchzer defended his curiosity, and his geographical inclusivity, by arguing that in order to understand the natural history of one region, one had to understand the natural history of all others, since they had all been created by God. "All of the lands of the world are joined together as if by a chain.... The natural history of one region is connected with that of others, even those a long way away; and one can hardly be discussed without that of the other.... Switzerland must often be sought in the Indies, and the Indies in Switzerland; this will be obvious from the book itself."[19] While the image that Scheuchzer used to justify his theory of geographical interconnectedness, namely that of the chain,[20] bears obvious resemblance to that of the Great Chain of Being, it is worth noting here that Scheuchzer uses it to suggest a horizontal link between places as well as a vertical link among species. With his splendid and surprising conjunction of Switzerland and the Indies, rhetorically uniting two regions as distant and different from each other as possible, Scheuchzer argued that an understanding of the natural history of other areas was, in fact, quite relevant for the understanding of one's own. Drawing thus on the culture of comparison he and his predecessors had created, Scheuchzer pushed it to its logical conclusion: knowledge of the natural history of other areas was needed not just for comparison, but to tell one more about one's *own* native land, so integrally linked to these other areas. A whole world of reference was thereby opened up.

Scheuchzer's inclusive outlook showed up strikingly in the ways in which he chose to commend the individuals involved in local natural history in his *Bibliotheca*. Scheuchzer listed not only individuals who had published books on natural history, such as local floras or regional mineralogies, but also acknowledged those who had written isolated journal articles, or had even mentioned in letters to others that they had contemplated writing some kind

---

[18] Scheuchzer, *Bibliotheca*, sig.)(3v.   [19] Scheuchzer, *Bibliotheca*, sig.)(5v.
[20] See, of course, Arthur O. Lovejoy, *The Great Chain of Being* (Cambridge, MA: Harvard University Press, 1936).

of natural-historical work about their region.[21] In Scheuchzer's willingness to recognize all varieties of intellectual effort may be seen an acknowledgement of the multiple ways in which individuals had contributed to natural history. Scheuchzer stressed that "in the study of nature you need experiments, *autopsia* (seeing for yourself), and journeys, but you also need Libraries, and you need the labors of others, however small and humble they may be...."[22] In recognizing the value of "small and humble" labors, Scheuchzer presented natural history as a collaborative effort, involving many different kinds of contributions, and potentially many different kinds of contributors.[23]

Scheuchzer's stance of openness may also be seen, for example, in the way he treated those whose contributions to natural history lay outside of the written word. For Scheuchzer, publication was not the universal prerequisite of scholarship, though it was certainly much to be desired; he listed numerous compilations of local natural history existing in manuscript ("MSC.") form. By doing so, he provided an incentive for interested readers to consult these, and to incite their authors to these works' completion.[24] Scheuchzer repeatedly cited manuscripts as in the process of emergence, listing authors as "preparing" their manuscripts (presumably for publication, but also possibly for circulation in the scribal tradition).[25] Nor did Scheuchzer limit his implicit definition of creativity to the act of writing, or to verbal expression, but commented favorably on those who had drawn or

---

[21] To cite just one of many possible examples, on one occasion Scheuchzer listed a certain "Goisson, a medical doctor at Lyons," as "working on a history of plants growing around Lyons." Even though this work was clearly still in progress, and had not yet even reached a stage at which it might be circulated, whether in manuscript or printed form, Scheuchzer still found it worthy of note. See Scheuchzer, *Bibliotheca*, 11.

[22] Scheuchzer, *Bibliotheca*, sig. [)(6]r.

[23] For a thoughtful discussion of the problems of coordinating similar volunteer networks of observers in a later period, see Katherine Anderson, *Predicting the Weather: Victorians and the Science of Meteorology* (Chicago, IL: University of Chicago Press, 2005).

[24] Scheuchzer, *Bibliotheca*, 19, 23, 44–45, 99, 104, 110–111, 146, 155, 161, 168, 173, 181, 189, and this is only a sampling. Scheuchzer listed even more of these "MSC." in his manuscript revisions to the *Bibliotheca*, while noting that others had in fact turned into published books in the meanwhile. For example, he listed the projector Urban Gottfried Bucher (discussed in the previous chapter) as working on a project with the title *Von Natürlicher Beschaffenheit und vermögenheit der Landgraffschaft Fürstenberg nebst einigen Oeconomischen reflexionen*, suggesting that Scheuchzer's extensive correspondence network had yielded him news of this as-yet-unfinished project; this title was subsequently crossed out and replaced with the book's ultimate title when published in 1720 (*Der Ursprung der Donau in der Landgraffschafft Fürstenberg*). See Zentralbibliothek Zürich, Ms. Z VIII 18, fol. 47r.

[25] Scheuchzer, *Bibliotheca*, 28, 61, 148. On the circulation of manuscripts in early modern Europe and, more generally, the persistence of manuscript culture in an "Age of Print," see Harold Love, *Scribal Publication in Seventeenth-Century England* (Oxford: Oxford University Press, 1993) and David McKitterick, *Print, Manuscript and the Search for Order, 1450–1830* (Cambridge: Cambridge University Press, 2003).

painted (*pinxit*) representations (*icones*).²⁶ By doing so, he drew on the entire inheritance of visual representation that had been of such crucial importance in defining the novelty of early modern natural history. As recent scholarship has noticed, many early modern illustrators of natural-historical and other works came from craft traditions, or were female; by taking these kinds of efforts into account, Scheuchzer reworked traditional definitions of scholarship.²⁷

Scheuchzer can also be seen as remarkably inclusive in the way in which he chose to list, alongside more learned works such as local floras, numerous popular sixteenth-century pamphlets about marvelous natural phenomena such as earthquakes or insect-showers. For example, Scheuchzer devoted an entire section at the end of his listing of works covering "Germania" to "anonymous Germans" (*Anonymi Germani*).²⁸ This section consisted entirely of pamphlets in the vernacular (mainly in German, with a few in Dutch), with such titles as *Wahrhaffte Zeitungen, die etlich 100. Menschen am Himmel zu Ingolstadt, zu Regenspurg und zu Nürnberg gesehen* (True Reports about what 100 People Saw in the Sky in Ingolstadt, Regensburg and Nuremberg, 1554) and *Ein wunderbahre Geburth und Veränderung der Natur von einer Hennen in Thoma* [sic] *Eberlins Burger zu Kolmar Hauss d. 18 Apr. 1538* (A Wonderful Birth and Change in the Nature of a Hen in the House of Thomas Eberlin, a Citizen of Colmar).²⁹ To some early Enlightenment observers, such works might well have seemed outdated or worse, reflective of popular superstition and the dangers of unguided speculation.³⁰ There is evidence that Scheuchzer himself shared similar beliefs. Yet the very fact that he chose to include these "anonymous" sixteenth-century pamphleteers testifies to his conviction that they too deserved at least some place in the history of natural history.³¹

This apparent attitude of inclusivity, furthermore, may also be seen in the way Scheuchzer chose to cite various non-European individuals who had collaborated in the production of books about their areas. This practice of citing non-Europeans, rarely carried out systematically at the time, appears most clearly in his section on the natural history of the Indies. Here Scheuchzer listed, along with the various European authors who had written

²⁶ Scheuchzer, *Bibliotheca*, 19, 149, 183.
²⁷ Londa Schiebinger, *The Mind Has No Sex? Women in the Origins of Modern Science* (Cambridge, MA: Harvard University Press, 1989), 66–79 and 102–104, discusses the contributions of women as artists and illustrators in the craft tradition.
²⁸ Scheuchzer, *Bibliotheca*, 66–81.   ²⁹ Scheuchzer, *Bibliotheca*, 66 and 68.
³⁰ See Katharine Park and Lorraine Daston, "Unnatural Conceptions: The Study of Monsters in Sixteenth- and Seventeenth-Century France and England," *Past and Present* 92 (1981): 20–54.
³¹ In his manuscript revisions to the *Bibliotheca*, Scheuchzer added still further examples of the pamphlet literature, including one, for example, on acid rain: Daniel Beckher's *Kurtzes und einfältiges Bedencken von des Schweffels-Regen* (Hamburg: Kleinhans, 1634). See Zentralbibliothek Zürich, Ms. Z VIII 18, fol. 40v.

books about the natural history of India, the names of those inhabitants of the subcontinent who had served as informants for Hendrik van Reede tot Draakestein, a Dutch governor of the province of Malabar, in his flora of that province.[32] So, for example, Scheuchzer listed "Apu Boto, one of the Brahmin Doctors from whom van Reede tot Draakestein took his knowledge of the plants of Malabar in the years 1674 and 1675."[33] As we can see here, Scheuchzer's focus was not solely on authorship, but on a broader kind of intellectual work. Thus, he also included in his bibliography entries for Itti Achudem, who "in the year 1670 also worked on the *Hortus malabaricus*,"[34] and for "Colladda. A Doctor of Malabar . . . who also put effort into producing the *Hortus Malabaricus*"[35]; and for several other native informants, alphabetized and mixed together with European authors.[36] In short, Scheuchzer's vision of natural history may be seen as having been inclusive not only in subject and scope, but also in its conception of those who counted as practitioners of natural history.

Scheuchzer's willingness to acknowledge local informants was certainly limited. It is worth noting, for example, that Scheuchzer stressed the *learned* qualifications of van Reede's informants Apu Boto, Itti Achudem, and Colladda, calling the first of these "one of the Brahmin Doctors" and the latter two simply "Doctors." While the degree to which the term of "Doctor" was intended to imply medical and/or learned status for the three remains unclear, Scheuchzer was probably aware that the term "Brahmin" in the first case was no random appellation, but rather an indication of honor and status

---

[32] Hendrik Adriaan van Reede tot Draakestein, *Hortus Indicus Malabaricus* (Amsterdam: sumptibus J. van Someren et J. van Dyck, 1678–1703). On this work, see J. Heniger, *Hendrik Adriaan van Reede tot Drakenstein (1636–1691) and Hortus Malabaricus: A Contribution to the History of Dutch Colonial Botany* (Rotterdam: Balkema, 1986); Marian Fournier, "Enterprise in Botany: Van Reede and his Hortus Malabaricus," *Archives of Natural History* 14 (1987): 123–158, 297–338; and K. S. Manilal, ed., *Botany and History of Hortus Malabaricus* (Rotterdam: Balkema, 1980). Scheuchzer was, it must be noted, only able to cite van Reede's informants because the latter had himself chosen to include this information in his own preface.

[33] Scheuchzer, *Bibliotheca*, 176.   [34] Scheuchzer, *Bibliotheca*, 182.

[35] Scheuchzer, *Bibliotheca*, 179. A fragment of a set of Scheuchzer's biographical notes survives in the British Library, unfortunately covering only the letters "Ca" through "Co"; here Scheuchzer gives still further details on Colladda, again suggesting, at the very least, a desire for thoroughness that worked to include, rather than exclude, information about informants. See British Library, Sloane mss. 3402, fol. 25r.

[36] See Scheuchzer, *Bibliotheca*, 186 for his citation of "Ranga Botto"; and 190 for his citation of "Vinique Pandito." It is also worth noting that Scheuchzer acknowledged several European collaborators of Reede's: see 178–179, and 183. In Scheuchzer's rough draft for his *Historiae Helveticae naturalis prolegomena*, we can see the origins of the process by which Scheuchzer decided to acknowledge these local informants. Scheuchzer had at first just written the title of van Reede's book on his list of natural histories of the East Indies; he subsequently added the names of the Indians themselves in the margins, presumably after looking more deeply into the *Hortus Malabaricus*, and this addition made it into the printed version of the *Prolegomena*. See Zentralbibliothek Zürich, Ms. Z VIII 601, fol. 14v.

in the Indian world.[37] It also remains unclear how many local informants Scheuchzer may have had the opportunity to cite, but did not. Certainly the people listed in his *Bibliotheca* remained predominantly learned men, with more than a fair share of academics among them; this might be contrasted, for example, with the even greater inclusiveness of the naturalist Maria Sibylla Merian around the same time, who repeatedly cited the Amerindian and Afro-Caribbean informants who had helped her with her natural-historical researches in Surinam.[38] In Scheuchzer's own previous and subsequent works on the natural history of Switzerland, in contrast, the majority of those he acknowledged were indeed scholars like himself, though with a considerable number of patricians and local notables as well. Even when Scheuchzer in other writings exhorted naturalists to gather information from farmers and the like, at least one motivation for his doing so, as we have seen, was to get local notables involved.[39] In his writings, it was then these gentry and their letters to him that he credited, not any oral communications between estate-owner and social subordinate that might have preceded this interchange.

But in order to understand Scheuchzer's attitude towards the individuals who helped create the works he discussed, it is perhaps best to turn to the passages in the *Bibliotheca* where Scheuchzer, before listing the works from a given region, briefly discussed the status of natural-historical knowledge of the region in question. Rather than recasting his information on this topic into the form of a map, as a modern scholar might do (see Figure 1 in the Introduction), Scheuchzer instead expressed his opinions in verbal form. In these passages, while quoting liberally from a popular early modern satirical compendium of national stereotypes,[40] Scheuchzer ended up generalizing about the quantity and quality of each region's contributions to

---

[37] In the case of Itti Achudem, for example, Scheuchzer, *Bibliotheca*, 182 seems to have been particularly impressed that he came from a family of "doctors," being "born to parents, grandparents, and great-grandparents who were physicians or doctors" (*Medicis sive Doctoribus*). Here it is clear, from his use of the word "physician," that Scheuchzer *does* see Itti Achudem more as an equivalent to a European physician.

[38] Natalie Zemon Davis, *Women on the* Margins (Cambridge, MA: Harvard University Press, 1995), 176–177, 184–187; Schiebinger, *Plants and Empire*, 35, 76, 206–207. On the issue of the acknowledgment (or lack thereof) of nonlearned informants and/or assistants more generally, see Anne Secord, "Artisan Botany," in *Cultures of Natural History*, ed. N. Jardine, J. A. Secord, and E. C. Spary (Cambridge: Cambridge University Press, 1996), 378–393, and Steven Shapin, "The Invisible Technician," *American Scientist* 77 (October 1989): 554–563.

[39] See Scheuchzer's exhortation of naturalists to include "curious men of the lowest order, fishers, shepherds of flocks, inhabitants of the Alps, farmers, diggers, root-cutters..." in his *Charta Invitatoria, Quaestionibus, quae Historiam Helvetiae Naturalem concernunt, praefixa* (Zürich, 1699), 2, discussed in the previous chapter.

[40] John Barclay's *Icon animorum*, originally published in London in 1614, which went through numerous editions in both Latin and assorted vernaculars over the course of the seventeenth century (and on into the eighteenth).

natural-historical knowledge. In the process, he elaborated what we might choose to term a geopolitics of science. Most importantly, he drew a set of clear distinctions between those areas which, he felt, had done a good job of documenting their own natural history, and those which had left the task to others, such as foreign travelers, to accomplish. Scheuchzer clearly saw the naturalists of central Europe as exemplary in the degree to which they had studied their own natural worlds.[41] Well over half of the book's space was enlisted in listing the productions of "Germania" and "Helvetia" (Switzerland), a clear imbalance which Scheuchzer, however, admitted might well be due to his own "ignorance" of other regions.[42]

In his discussions of the geography of natural history, Scheuchzer repeatedly seems to have operated on the assumption that the inhabitants of an area – what we might call its indigenes – not only were the ones most capable of reporting on its natural history, but also, as such, had a moral responsibility to share these reports with others elsewhere. We can see this, for example, in Scheuchzer's comments on those regions that he seems to have seen as being on the margins of Europe, such as Spain and Portugal on its western fringe, and Hungary and Poland on its eastern. Regarding Spain, Scheuchzer was at first appropriately apologetic about his skimpy citations of natural-historical discussions of the Iberian peninsula, noting prominently in his introduction itself, before even commencing his catalogue, that "You will find few Spanish in this society of learned people [i.e. his book]; do not attribute this to the negligence of this most polite nation, but to my own ignorance, and the still limited commerce of Europeans with the Spanish."[43] Here Scheuchzer does indeed seem to acknowledge his "own ignorance," suggesting that there may have been fundamental structural causes, for example in European trade patterns, for his own lack of knowledge about Spanish natural history.

Later on in the book, though, Scheuchzer revealed a more fundamental concern with questions of patriotism and authorship, noting with evident disapproval that Spanish plants had been "described more by foreigners, than by the inhabitants (*incolis*) themselves."[44] Scheuchzer does seem to have had a point, at least as far as the publication of printed works; indeed, far less material does seem to have been published on the natural history of the Iberian peninsula during this period than on other areas, and much

---

[41] Scheuchzer, *Bibliotheca*, 3–4 and 17.   [42] Scheuchzer, *Bibliotheca*, sig.)(3v.

[43] Scheuchzer, *Bibliotheca*, sig.)(3v. Note, though, that in this sentence, Scheuchzer presents "Europeans" and "Spanish" as mutually exclusive categories. On the other hand, his manuscript indices of natural-historical authors show the extent to which he had attempted to remedy his ignorance; for example, the very first entry in a manuscript "Bibliotheca botanica" he compiled (and presumably kept adding to as he worked) was a reference to a Spanish physician, namely a certain "Abdalla" of Toledo who in the High Middle Ages had written on agriculture. Zentralbibliothek Zürich, Ms. Z VIII 611, fol. 25r.

[44] Scheuchzer, *Bibliotheca*, 4.

of it was indeed published by travelers from elsewhere in Europe.[45] But the word "published" here is key. Recent scholarship on the prodigious (and often unpublished) researches conducted by Spanish writers on the flora and fauna of the New World suggests that there was, in fact, far from a shortage of information-gathering on the natural world; but that it was directed primarily at the Americas, rather than Europe, and that far from wanting to make this politically and economically sensitive information publicly available, Spanish imperial officials much preferred to carefully guard it and keep it secret from potential commercial interlopers.[46] Scheuchzer may or may not have been aware of some aspects of this situation; in any case, he certainly frowned on its consequence, a lack of publicly accessible information on Spanish natural history. Comparing Spain with Germany, Scheuchzer suggested that "Whereas the German nation appears to err in excess, in the disease of writing too much, the Spanish err in the defect thereof."[47] For Scheuchzer, the fact that naturalists elsewhere in Europe had to turn to accounts written by the Belgian-born botanist Clusius and by the English-born Grisley for their knowledge of the Iberian peninsula was indeed one worthy of comment. The Spanish, he seems to have insinuated, had neglected their patriotic duty to write about their own "fatherland." Having neglected this duty, they had left for others the work that they should have done themselves.

Scheuchzer was thus keenly interested in questions of the *origins* and authorship of the natural-historical works he investigated. In the case of Hungary, he charitably ascribed Magyar deficiencies in inventorying their lands to external circumstances (namely war with the Turks): "The nation

[45] Of the three works Scheuchzer cites as discussing Spanish natural history (an incredibly small number compared with the dozens of works he lists for every other part of Europe except Eastern Europe), two of the authors involved were English, and the third Flemish: Carolus Clusius, *Rariorum aliquot stirpium per Hispanias observatarum historia* (Antwerp: ex officina Christophori Plantini, 1576); Gabriel Grisley, *Viridarium lusitanicum* (Lisbon: ex praelo Antonii Craesbeeck, 1661); and John Polus Lecaan, *Advice to the Gentlemen in the Army of Her Majesty's Forces in Spain and Portugal: with a short Method how to preserve their Health . . . To which are added the Medicinal Virtues of many peculiar Plants* (London: Printed for P. Varenne, 1708). On the latter, see Maria Antònia Martí y Escayol, "Catalunya dins la xarxa científica de la il·lustració. John Polus Lecaan: medicina i botànica a Barcelona durant la Guerra de Successió," *Manuscrits* 19 (2001): 175–194.
[46] See Antonello Gerbi, *Nature in the New World: From Christopher Columbus to Gonzalo Fernandez de Oviedo* (Pittsburgh, PA: University of Pittsburgh Press, 1985); Raquel Álvarez Peláez, *La conquista de la naturaleza americana* (Madrid: Consejo Superior de Investigaciones Científicas, 1993); Daniela Bleichmar, "Books, Bodies, and Fields: Sixteenth-Century Transatlantic Encounters with New World *Materia Medica*," in *Colonial Botany: Science, Commerce, and Politics in the Early Modern World*, ed. Londa Schiebinger and Claudia Swan (Philadelphia: University of Pennsylvania Press, 2005), pp. 83–99; and Antonio Barrera-Osorio, *Experiencing Nature: The Spanish American Empire and the Early Scientific Revolution* (Austin: University of Texas Press, 2006).
[47] Scheuchzer, *Bibliotheca*, 3–4.

of the Hungarians is a warlike one, and with an infestation of constant wars and persecutions, it does not appear they have enough leisure to be able to devote any time to natural-historical study. [The writings] which exist on this subject are extremely few, and almost none of them are by Hungarians themselves."[48] Scheuchzer was less charitable towards the Poles. In his opinion, Poland was hampered by a "harsh life, bitter weather, and customs of the people not fashioned according to the politeness of our age." The result, he opined, was that "a barbarity of customs brings with it a barbarity of studies," leading again to a paucity of writings on natural history by the inhabitants (*inquilini*) themselves.[49] Regardless of the causes Scheuchzer saw as responsible for the insufficient documentation concerning certain areas, he was clearly distressed by this gap in his chain of knowledge; and he chose to express this by lamenting the lack of truly local input and agency, namely that of the inhabitants themselves, in the existing literature.

When Scheuchzer's attention turned to the natural history of Asia, Africa, and the Americas, he likewise noted that the inhabitants of these continents had not "produced" much natural history, but here seems to have resigned himself to the thought that "Europeans" would indeed have to be relied upon for further accounts of these regions. For example, Scheuchzer held that Asia had once been the "domicile of the Muses," but had slipped from its formerly civilized state into a state of "sad ignorance," to the extent that "whatever has been brought forth about the natural history of those areas, is owed to Europeans."[50] Scheuchzer sardonically pointed out that this knowledge garnered by Europeans was not, however, the result so much of interest in natural history *per se*, as rather of European greed for the spice trade and other sources of potential riches: "And those [Europeans] would never have thought as much about advancing this study [natural history], if the holy hunger for gold (*auri sacra Fames*) hadn't drawn them there."[51] In discussing each individual Asian nation, whether China, Japan, or "Turkey, Tartary, the Empire of the Great Mongol, and that of the Persians," Scheuchzer returned again and again to the same theme, that "we don't know about what is being cultivated there, except from the journeys that Europeans have taken up there...."[52] In Africa and the Americas, we see Scheuchzer stressing this interpretation yet again, that the inhabitants of these regions had abandoned their duties, and that the task had subsequently been taken over by Europeans.[53] Here, however, Scheuchzer gave up all pretense of urging these regions' inhabitants to cultivate their own talents, but rather urged Europeans to go ahead and

[48] Scheuchzer, *Bibliotheca*, 140.
[49] Scheuchzer, *Bibliotheca*, 142. On changing Western & Central European attitudes towards Eastern Europe at the time, see Larry Wolff, *Inventing Eastern Europe: The Map of Civilization on the Mind of the Enlightenment* (Stanford, CA: Stanford University Press, 1995).
[50] Scheuchzer, *Bibliotheca*, 174-175.   [51] Scheuchzer, *Bibliotheca*, 175.
[52] Scheuchzer, *Bibliotheca*, 175.   [53] Scheuchzer, *Bibliotheca*, 192 and 198.

pursue the natural history of, in the latter case, "the vastness and fertility of this New World."[54]

In his *Bibliotheca*, Scheuchzer thus developed a complex vision of the usefulness of local natural history to the natural-historical enterprise. Again and again he stressed that knowledge of a wide variety of regions, acquired from a wide range of individuals and sources, was crucial to natural-historical understanding more broadly, as well as to the understanding of Switzerland's own natural world. Yet Scheuchzer's vision was also far from straightforward. His geographical methods of organizing his subject-matter ended up stereotyping the lands and peoples he discussed, as much as acknowledging their potential. For Scheuchzer, the demands of natural history on its followers were so great that all possible collaborators, he believed, must be enlisted, regardless of their flaws. For Scheuchzer, natural history was not a centralized enterprise, but rather a dispersed one, spread out over many lands.

FLORISTS AND CRITICS

Linnaeus's *Bibliotheca botanica* or "botanical library," published in 1736, provided a very different view of the role and importance of local knowledge. Whereas Scheuchzer had reveled in the rich variety of localist works, Linnaeus, with his famous new system of classification, found their very diversity frustrating, and sought to discipline them. Through his new system of binomial nomenclature, by which he assigned a single genus and species to every living entity, he notoriously reduced much botanical work to the counting of plants, pistils, and stamens. In this section as in the previous one, the purpose will be to examine Linnaeus's *Bibliotheca* not so much to shed light on an individual's career, as rather to probe the ways in which Linnaeus's *Bibliotheca* represents a different approach to the problem of local knowledge than that presented by Scheuchzer. In his bibliography of botany, I will argue, Linnaeus was to present a system of natural-historical

---

[54] Scheuchzer, *Bibliotheca*, 198. Indeed, over the course of the eighteenth century, Europeans seem to have accepted Scheuchzer's advice, whether conscious of it or not; the century came to be seen as the great Age of Exploration, with yet a new generation of European heroic voyagers exploring the Pacific as well as the Atlantic. For recent interpretations of these voyages of exploration, see for example David Philip Miller and Peter Hanns Reill, eds., *Visions of Empire: Voyages, Botany, and Representations of Nature* (Cambridge: Cambridge University Press, 1996); Mary Terrall, "Heroic Narratives of Quest and Discovery," *Configurations* 6 (1998): 223–242; Dorinda Outram, "New Spaces in Natural History," in *Cultures of Natural History*, ed. N. Jardine, J. A. Secord, and E. C. Spary (Cambridge: Cambridge University Press, 1996), 249–265; and her "On Being Perseus: New Knowledge, Dislocation, and Enlightenment Exploration," in *Geography and Enlightenment*, ed. David N. Livingstone and Charles W. J. Withers (Chicago, IL: University of Chicago Press, 1999), 281–294.

knowledge which sharply devalued the contributions of local natural history, and of certain kinds of local informants.

Yet in fact, despite his later reputation as a botanical imperialist, Linnaeus got his start in much the same kind of world of local natural history as that we have explored, and many of his ideas came from within it. Trained at the University of Uppsala in the 1720s, he participated along with all his fellow medical students in the same botanical excursions, and was fully versed in the necessity for *autopsia* and for the direct investigation of local nature. One of his first works was a diary of a journey into Lapland, in which he sought out the plants and other natural features of this as yet unstudied region of Sweden.[55] For him, the *patria* or native land was the necessary foundation for all natural history, and in his mature years, as professor at Uppsala, he was to insist that his students acquire the same direct contact with nature, through botanical excursions, that he himself had gained.

An address Linnaeus delivered in 1741, urging the necessity of "travelling inside one's own native land," provides particularly telling evidence of Linnaeus's localist convictions.[56] In this speech, which was published at the time in Uppsala and, two years later, in the Netherlands, Linnaeus argued that too many Swedish youths were heading off on European tours, without having first fully explored what their *patria* had to offer. "Good Lord! how many people, ignorant of the affairs of their native land, rush to foreign parts, so they can examine and admire the curious things offered up to them there; of which the greatest part are far below the level of those which, in our native soil, force themselves on the attention of those who only open their eyes. I have not, while abroad, seen any soil more fertile in curious objects belonging to the kingdoms of nature, than that of our native land."[57] Inserting example after example of Sweden's own natural wonders, and drawing lavishly on his own Lapland adventures in doing so, Linnaeus promised his listeners that if they only thoroughly investigated what their own native land had to offer, they would "... discover such things, as by their own admission, they would never have even been

---

[55] See Carl Linnaeus, *The Lapland Journey*, edited and translated by Peter Graves (Edinburgh: Lockharten Press, 1995); this work remained unpublished during his life, while he converted his notes into a local flora of Lapland. On Linnaeus, see also James Larson, *Reason and Experience: The Representation of Natural Order in the Work of Carl von Linné* (Berkeley: University of California Press, 1971); Lisbet Koerner, *Linnaeus: Nature and Nation* (Cambridge, MA: Harvard University Press, 1999); and Staffan Müller-Wille, *Botanik und weltweiter Handel. Zur Begründung eines natürlichen Systems der Pflanzen durch Carl von Linné (1707–1778)* (Berlin: VWB, 1999).

[56] Linnaeus, *Oratio qua peregrinationum intra patriam asseritur necessitas* (Uppsala, 1741). This address was also later published abroad as *Oratio de necessitate peregrinationum intra patriam* (Leiden: apud Cornelium Haak, 1743); I have used this latter version. All translations are my own; an eighteenth-century English translation does exist, namely in Benjamin Stillingfleet's *Miscellaneous Tracts*, but it differs in some respects from the original.

[57] Linnaeus, *Oratio*, 11.

able to imagine in their dreams."[58] Furthermore, the things they found in their *patria* would "not only nourish and satisfy their desire for knowledge, but also enable them to serve their native land, themselves, and the public good."[59] Linnaeus thus integrated the activity of local travel into a cameralist framework, one capable of producing benefits to individual and community alike.

But Linnaeus's intentions must be examined more closely. Despite his enthusiasm about the advantages that might accrue to the "public good" from traveling within the native land, Linnaeus stressed that such travel was not, in fact, for everybody. "Even though it is necessary, as we have already mentioned, to investigate our native soil, anyone who takes on this labor will do so uselessly and in vain, if he has not previously learned the fundamentals of his studies in the Academies, in *Physics, Natural History* and *Medicine* [emphasis in original]."[60] Only those with suitable academic preparation could possess the necessary "attention" to detail, that is to say, only students need apply. In any case, students needed to practice on local settings before venturing out into the foreign world. In short, Linnaeus designed a kind of local travel that would be restricted to students, as a preliminary exercise preparatory to further assignments in natural history.

Linnaeus asserted that the kind of travels of which he spoke had hardly been attempted in the past. He spoke, he said, "not about the advantages of academies in the native land, nor about the advantages of travelling to these, but about instituting travels (*peregrinationibus*) within the native land, through its fields and crossroads; about that type of travels, which up to this day, I admit, have been less in use, or which have been believed to have been purely for amusement."[61] This interpretation might be read in several ways. First, was Linnaeus really trying to claim that local travel had, prior to his own efforts, been undervalued? As the preceding chapters have shown, this was certainly not the case in the early modern German territories, where many naturalists had engaged in exactly these kinds of activities, using many of the same arguments. Nor had Sweden remained free of this influence; local floras had been written of university towns such as Uppsala, and other investigations to this end attempted. Another way of reading this statement might be to see Linnaeus as asserting that it was his techniques that were new, that his insistence on close observation had lent natural-historical travel a new dimension. Here again, however, it appears that Linnaeus was not particularly unique, but rather one of a number of naturalist reformers around this time making such claims.

Linnaeus, then, was very much a product of contemporary interests in local natural history. Yet in his *Bibliotheca botanica*, he was to assign local

[58] Linnaeus, *Oratio*, 12.   [59] Linnaeus, *Oratio*, 12.   [60] Linnaeus, *Oratio*, 23.
[61] Linnaeus, *Oratio*, 10.

knowledge a very different role than the prominent one which Scheuchzer had given it. Whereas Scheuchzer had devoted his entire catalogue to "special" natural history, Linnaeus produced in his work a taxonomy of botanists which reduced "florists" to one single category out of sixteen. In this classification of botanists, authors of local floras shared space with "fathers of botany," "commentators," "illustrators," "describers," "writers of monographs," "curious men," "garden-lovers," "travelers," "anatomists," "horticulturalists," "medical doctors," and "anomalous."[62] In Linnaeus's view, all of these shared the common feature of being either "collectors" or mere "lovers of botany" (*botanophili*), a term which he explicitly counterposed to "botanists" (*botanici*). In contrast, Linnaeus set himself among the botanists and, in particular, among what he called the "methodists" (*methodici*), including "philosophers," "systematists," and the "namers of things."[63] Linnaeus founded his classification of botanists upon an analysis of the kinds of activity they performed, whether that of drawing, or traveling, or simply thinking; and he ranked these activities in a hierarchy of botanical performance, with taxonomists at the top and florists occupying a decidedly inferior position.

Linnaeus's attitude towards the florists was critical, to say the least. He noted that up until the present date florists had used a disturbing variety of synonyms in the titles of their works, for example not only "flora" but also "index," "enumeration," "catalogue," "synopsis," "mirror," "vademecum," and so forth; he recommended "flora" as a universal substitute.[64] In his view, this unsettling diversity of titles reflected an even greater problem with their contents. He complained that

all who have written floras of these kinds have made up names of their own for plants, and without the testimony of authors; and they wrote them down in their own ways, that they themselves knew, and they didn't really understand other authors, whence the fact that nobody can understand them; if somebody in our time wanted to write in such a method, I would urge him in a friendly manner that he desist from such a vain work, so that no-one might complain of his ignorance.[65]

What Linnaeus really disliked in the past century's local floras, as we see from this extremely revealing passage, was their methodological eclecticism, the way they often listed multiple names for the same plant, and sometimes even invented their own. In short, he believed that the authors of local floras had descended too far into the local; that their methods, as well as their objects of study, had become localized to an extent which seriously impeded

---

[62] Carolus Linnaeus, *Bibliotheca botanica* (Amsterdam: apud Salomonem Schouten, 1736), sig. [*7]v-[*8]v.

[63] Linnaeus, *Bibliotheca*, sig. [*7]v and [*8]v.   [64] Linnaeus, *Bibliotheca*, 84.

[65] Linnaeus, *Bibliotheca*, 84.

botanical communication – that their works had become almost mutually incomprehensible. Linnaeus proposed that "it would be to be hoped that all florists should write systematically, and that the only people who should write local floras should be those who are more learned."[66] Providing examples of a select few who had "written their floras systematically," he urged that *their* examples be followed.[67] In this way, Linnaeus pushed for a reorganization of the relations between the local and global, that would enable the universalization of botanical knowledge.

Linnaeus was not mean-spirited in his disparagement of the local florists. The change he was advocating was a structural one, that was intended to advance all of the different categories of botany he discussed in the *Bibliotheca*, far more than to denigrate any particular one. Furthermore, Linnaeus actually envisioned a more active role, in many ways, for such individuals as women and peasants, whom he encouraged to take part in gathering local plants.[68] And in many ways Linnaeus can be seen as encouraging his students to pay critical attention to local knowledge, particularly when it concerned uses of plants.[69] Nonetheless, Linnaeus's censure of the local flora reflects a broader theme in his works, namely that of the advancement of what we might term "theory" within botany. One of Linnaeus's chief aims, most of his biographers have agreed, was to enable scholarly botanists at the best botanical gardens in the world, surrounded by their vast collections of species from all over the globe, to create new visions of how nature might be successfully ordered. To realize this vision, local knowledge had to be standardized, and its scattered collectors disciplined so that they would all follow a single method. From henceforth, while their labors would be recognized as useful and certainly crucial to the continuation of the botanical enterprise, naturalists on the margins would tend to be relegated to the sphere of "practice," while others in such centers as Paris and London would make theory, and thereby make history.[70] Though local floras, and other works concerning the local natural history of specific regions, would continue to be written in ever increasing quantities, their place within the economy of scientific knowledge would never be the same.

The 1730s marked the end of an era in natural history. Though the triumph of the Linnaean system was to be a longer and more complex story than is often recognized, and though it in itself was to be only the beginning of a set of further developments in the systematic understanding of the natural

[66] Linnaeus, *Bibliotheca*, 85.   [67] Linnaeus, *Bibliotheca*, 85.
[68] Lisbet Koerner, "Women and Utility in Enlightenment Science," *Configurations* 2 (1995): 233–255.
[69] Lisbet Rausing, "Underwriting the Oeconomy: Linnaeus on Nature and Mind," in *Oeconomies in the Age of Newton*, ed. Margaret Schabas and Neil De Marchi (Durham, NC: Duke University Press, 2003), 173–203, especially 198.
[70] Schiebinger, *Plants and Empire*.

world, crucial changes were already underway for authors of local floras and other works on local natural history.[71] Increasingly, the task became to revise previously existing works in accordance with Linnaean method, and this is how local floras and similar works now came to announce their goals. A set of social changes accompanied this shift. Students came to unite and publish local floras and the like as preliminary exercises establishing their credentials, before moving on to other and more exciting projects having little to do with local study. Individuals who devoted significant portions of their adult lives to the study of the natural history of their homelands or adopted lands, as Scheuchzer had, became increasingly rare. The Linnaean method, by simplifying the process of writing local floras and similar works, relegated them more and more to mere mechanical exercises, allowing no such liberties of method and description as their predecessors had. Whereas Scheuchzer had tended to assume that locals would usually be better informed about the natural history of their areas than others, the same was not true for the Linnaeans, for whom observers were interchangeable, as long as they possessed the relevant standardized skills. Linnaeus has recently been described as having trained his pupils in the art of alienation, of systematically *distancing* themselves from their own local surroundings by their cultivation of new techniques of observation.[72] Seen in this light, the Linnaean movement represented a move away from the kinds of truly place-based knowledge that early modern local natural history had represented.

In many ways, though, it might be more profitable to view the Linnaean movement as the culmination of a lengthy set of developments, visible even in the earliest local floras. In the ways in which they reduced their scope to the listing of plant names and places, local floras had themselves contributed in many ways to the effacement of local informants, and to the erasure of many of the kinds of folk and popular lore that had graced the herbals. Despite the advent of place as a central organizing category in the local floras, the knowledge possessed by the *non*-learned individuals who lived in these places had not been furthered; rather, it had experienced a quiet eclipse. In this way, much of what has been seen as Enlightenment disciplining of local knowledges may be seen as having occurred long before, in the gradual

[71] For the spreading of the Linnaean method, see Frans A. Stafleu, *Linnaeus and the Linnaeans: The Spreading of their Ideas in Systematic Botany, 1735–1789* (Utrecht: Oosthoek, 1971). For subsequent developments during the eighteenth century, see G. Baehni, "Les grands systèmes botaniques depuis Linné," *Gesnerus* 14 (1957): 83–93; James L. Larson, *Interpreting Nature: The Science of Living Form from Linnaeus to Kant* (Baltimore, MD: Johns Hopkins University Press, 1994); and Peter F. Stevens, *The Development of Biological Systematics: Antoine-Laurent de Jussieu, Nature, and the Natural System* (New York: Columbia University Press, 1994).
[72] Lisbet Koerner, "Daedalus Hyperboreus: Baltic Natural History and Mineralogy in the Enlightenment," in *The Sciences in Enlightened Europe*, ed. William Clark, Jan Golinski, and Simon Schaffer (Chicago, IL: University of Chicago Press, 1999), 419–420.

emergence of the very format and tradition of the local flora itself. While regional mineralogies and natural histories of individual territories offered alternative models of how to describe the natural world, it was in the end the model of the local flora, as Linnaeus adopted and adapted it, that may be seen to have prevailed.

Conclusion

In February 1775, a curious article entitled "Substitutes for Tea" appeared in *The Pennsylvania Magazine; or, American Monthly Museum*, a periodical published in Philadelphia. Penned by an anonymous author, who referred to himself only as "a Philanthropist," the piece commenced, innocently enough, with a brief commodity history of tea, that soon turned, however, into a harangue: "India Teas were first introduced into Europe A. D. 1679, by the extravagant encomiums of Cornelius Bontekoe, a Dutch physician: The tyranny of fashion spread it with amazing rapidity; the general state of health has undergone a great revolution by it, so that our race is dwindled, and become puny, weak and disordered, to such a degree, that were it to prevail a century more, we should be reduced to mere pigmies."[1] Citing medical authors aplenty who had condemned the "destructive exotic" for its pernicious bodily effects, and entire regions of Europe that had supposedly banned it, the anonymous author suggested that it was high time that substitutes be found for the offending plant. "But if we must, through custom, have some warm tea, once or twice a day, why may we not exchange this slow poison, which besides its other evils destroys our constitution, and drains our country of many thousand pounds a year, for teas of our own American plants: many of which may be found, pleasant to the taste and very salutary..."[2]

Leaping into the breach, the nameless "Philanthropist" went on to suggest no less than seventeen such substitutes for Indian tea, from sassafras and sweet marjoram to mistletoe and wild valerian (not counting warm water which, he assured his readers, was just as good!).[3] Serving so many different kinds of tea would, of course, cause extreme overloading of the tea-table, he commented; but, he insisted, it would be worth it, because "If the

---

[1] "Philanthropist," "Substitutes for Tea," *The Pennsylvania Magazine; or, American Monthly Museum* (February 1775), 73–74.

[2] "Philanthropist," 73, 75.

[3] "Philanthropist," 75. In the listings of substitutes, the anonymous author heartily mingled American native herbs with successful European transplants, perhaps in the recognition that his audience consisted in the main of European transplants themselves.

gentlemen and ladies of the first rank, will use their influence and example, to abolish this pernicious custom of drinking the Asiatic teas, and introduce and persevere in using our own; they will have the self-pleasing satisfaction of having emancipated their country from the tyranny and slavery of an evil custom, and erecting a monument to common sense, which will merit the praises of unborn generations."[4] Only by rejecting the exotic and seeking out the indigenous could moral virtue be maintained, and liberty from foreign corruption achieved.

The author of this odd little article may, in fact, very well have been Thomas Paine, the British-born American revolutionary, who was the editor of the *Pennsylvania Magazine* at the time. Or it may not. In any case, when the anonymous author wrote it, he (or she) was doing so within a very specific context. Slightly over a year previously, a band of Bostonians, disguised as Mohawk Indians, had boarded three tea-laden East India ships, the *Dartmouth*, the *Eleanor*, and the *Beaver*, and had dumped the fragrant cargo overboard, all forty-five tons of it. They had not done so, of course, out of medical motives. If the patriots of colonial New England were indeed to have diagnosed themselves as suffering from any particular disease, it would have been that of British tyranny – symptomatized by the Tea Act of 1773, imposed by Parliament to save the East India Company from bankruptcy, and to further punish the already rebellious Americans – and the only cure they would have admitted would have been liberation from that foreign tyranny.[5] For the colonists, in short, India tea stood proxy for British rule more generally. By symbolically (and physically) acting to prevent the unloading of this foreign commodity onto their wharves, the band of Bostonian "patriots" acted in a way that set a Revolution in motion.

Clearly, the rhetoric of debate over the "indigenous" had survived its journey across the Atlantic Ocean, and remained available to all in the late eighteenth century who wanted to adapt its imaginative geography to their own circumstances. Indeed, it survives and flourishes to this day, though the word "indigenous" is nowadays far more likely to be found referring to peoples and cultures *beyond* the European continent (or on its fringes) than within it.[6] Indigenous peoples from Canada to the Philippines, for example, as well as those on the margins of Europe, struggle today to recover their

---

[4] "Philanthropist," 76.

[5] On the Boston Tea Party and the American Revolution from the perspective of commodity history, see T. H. Breen, *The Marketplace of Revolution: How Consumer Politics Shaped American Independence* (Oxford: Oxford University Press, 2004), 294–331.

[6] The increasing 20th-century popularity of the word "indigenous" to refer to First Nations peoples may well have been an artifact of the 19th-century British imperial use of the term "native(s)," usually then applied to subject races overseas, which led inexorably to pejorative connotations and the subsequent abandonment of that word in many, though not all, connections.

rights, and in the process to right wrongs first committed hundreds of years ago, in many cases, during the very period this book has been discussing.[7] Simultaneously, they work to attempt to preserve indigenous languages and cultures from the pressures of contemporary globalization.[8] Among other groups of people who would not dream of calling themselves "indigenous," meanwhile, new forms of nativism arise, with attacks on immigrants and "foreigners" perceived as threatening the economic and social well-being of the "native-born" – often, as in the Americas, descendants of immigrants themselves.

Concerns over the proper relationship of the "indigenous" and the "exotic" are also stronger than ever, meanwhile, among those who study the natural world. Ecologists and conservation biologists have recently come to pay considerable attention to the ways in which particularly adaptive forms of exotics, now termed "invasive species," have been mounting "bioinvasions" and posing an increasing threat to those remaining areas of the globe that are still, five centuries after the initiation of the Columbian Exchange, predominantly inhabited by indigenous flora and fauna.[9] As rates of species extinction climb to new heights in what has been termed the "sixth great extinction," environmentalists today are now increasingly engaged in new strategies of conservation: in particular, that of involving indigenous *peoples* in the protection of indigenous species, with the future of those species now

[7] More and more is being written nowadays on this topic; see for an introduction Ken S. Coates, *A Global History of Indigenous Peoples: Struggle and Survival* (Houndmills, UK: Palgrave Macmillan, 2004). Within Europe, the Sami people of northern Sweden, Finland, Norway, and western Russia are generally seen as most clearly belonging to the category of indigenous peoples, though a number of other groups are beginning to lay claim to that same appellation; see for example the "Indigenous node" of the World Wide Web Virtual Library at www.cwis.org/wwwvl/indig-vl.html.

[8] On threats to indigenous languages, see Daniel Nettle and Suzanne Romaine, *Vanishing Voices: The Extinction of the World's Languages* (Oxford: Oxford University Press, 2000).

[9] On the history and continuing consequences of biological exchange, see for example Chris Bright, *Life Out of Bounds: Bioinvasion in a Borderless World* (New York: W.W. Norton, 1998); Robert S. Devine, *Alien Invasion: America's Battle with Non-Native Animals and Plants* (Washington, DC: National Geographic Society, 1998); Kim Todd, *Tinkering With Eden: A Natural History of Exotics in America* (New York: W. W. Norton, 2001); Peter Coates, *American Perceptions of Immigrant and Invasive Species: Strangers on the Land* (Berkeley: University of California Press, forthcoming); Gert Groening and Wolschke-Bulmahn, "Some Notes on the Mania for Native Plants in Germany," *Landscape Journal* 11, 2 (1992): 116–126; and the articles in Marcus Hall and Peter Coates, eds., "The Native, Naturalized, and Exotic: Plants and Animals in Human History," special issue of *Landscape Research* 28, 1 (2003). For a recent discussion of the 19th-century origins of terms like "bioinvasion," see James Moore, "Revolution of the Space Invaders: Darwin and Wallace on the Geography of Life," in *Geography and Revolution*, eds. David N. Livingstone and Charles W. J. Withers (Chicago, IL: University of Chicago Press, 2005), 106–132.

seen as inextricably intertwined with that of the peoples who have lived amongst them the longest.[10] In short, though the geographical contexts in which the "indigenous" is most commonly invoked may have shifted drastically since the early modern period, *concern* about issues relating to indigenous cultural and natural worlds may perhaps be greater than ever.

Where, then, does that leave this book? *Inventing the Indigenous* has sought to show the ways in which Europeans, in the wake of the Columbian encounter, came to understand their own "indigenous" natural worlds. Over the course of the early modern period, numerous factors joined together – cultural, medical, political, religious, institutional, economic – to spur Europeans to investigate the contents of their own local landscapes. It was amidst this quest to recover what Europeans saw as their own kinds of "indigenous" natural variety that detailed inventories of local nature began to be compiled and published. In this way, new genres for documenting the natural world came into being, from the local flora to the regional mineralogy and the natural history of the territory. And as a result, what we might call a self-consciously "local" knowledge of the natural world came to be formed in early modern Europe.

This "local" knowledge was profoundly shaped by the geographical contexts in which it was produced. As this book has shown, the earliest genre of local natural history, the local flora, emerged within the scattered towns and territories of the early modern Holy Roman Empire. These are not the kinds of sites that scholars have generally seen as the most conducive to the formation of new kinds of natural inquiry; with very few exceptions, most historians of seventeenth-century science and culture have preferred to focus on the seemingly more dynamic spaces of Western European countries like England and France, where, amidst monarchical centralization and burgeoning colonial movements, the earliest national scientific academies began to emerge.[11] Yet in the German territories, political and religious fragmentation led to the proliferation of various institutions, such as the university, that were to prove crucial in stimulating natural inquiry, as well as in

---

[10] See for example Virginia D. Nazarea, *Cultural Memory and Biodiversity* (Tucson: University of Arizona Press, 1998) and Luisa Maffi, ed., *On Biocultural Diversity: Linking Language, Knowledge, and the Environment* (Washington, DC: Smithsonian Institution Press, 2001).

[11] Exceptions include Bruce Moran, *The Alchemical World of the German Court: Occult Philosophy and Chemical Medicine of the Circle of Moritz of Hessen (1572–1632)* (Stuttgart: Steiner, 1991) and Pamela H. Smith, *The Business of Alchemy: Science and Culture in the Holy Roman Empire* (Princeton, NJ: Princeton University Press, 1994), both of whom show (in different ways) the very real preoccupations with natural knowledge that arose out of the practice of alchemy, traditionally dismissed by historians as an utterly "occult" activity. On stereotypes of early modern residents of the Holy Roman Empire as preoccupied with religious and mystical inquiries, see William Clark, "The Scientific Revolution in the German Nations," in *The Scientific Revolution in National Context*, eds. Roy Porter and Mikuláš Teich (Cambridge: Cambridge University Press, 1992), 90–114.

shaping the ways in which that inquiry came to be documented. Local floras came to be produced in so many German university towns precisely because they filled a very real need, among multiple medical faculties, for pocket-sized books that students could take with them on local botanizing field trips. Thus the very provinciality of so many of the small German towns in which universities were located helped to spur the creation of this new kind of "local" knowledge.

Likewise, when regional mineralogies began to be written in the second half of the seventeenth century, they too first appeared in the German territories. Here as well geographical factors played a prominent role, but in a different way than had been the case with the local flora. For regional mineralogies arose within areas that had already developed strong mining traditions, owing to their mountainous locations characterized by a high presence of concentrated metallic ores and other striking geological features. The distribution of such areas across Europe was naturally uneven; a very high percentage happened to be located within the boundaries of the Holy Roman Empire in areas like the Harz Mountains and the Erzgebirge. Authors of regional mineralogies took advantage of this fact, drawing on centuries of previous mining literature to produce their own far more thorough catalogs of the mineral productions of their areas. The works they generated differed greatly from the local floras they so often cited. Unlike those terse and tiny plant-lists, regional mineralogies were verbose compendia of all the practical and symbolic uses that could possibly be imagined for rocks and figured stones. They were thus well-suited to the needs of the audiences they aimed at, more often ambitious functionaries at small German princely courts than medical students. Here too, conditions within the Holy Roman Empire can be seen as playing a crucial role in shaping the ultimate form of the knowledge thus produced.

The emergence of the county natural history in 1660s England, though, posed a challenge to the forms of inventory of local nature that had developed in the German territories. Whereas local floras and regional mineralogies produced within the Empire had generally confined themselves to the documentation of their respective natural objects, eschewing antiquarianism, the county natural histories of England almost invariably did the opposite, offering their readers extensive accounts of local history, customs, and monuments. These differences in style meant that the new, English way of doing local natural history did not travel easily to the Continent. As we have seen, Royal Society intelligencer Henry Oldenburg attempted repeatedly to get his imperial correspondents to write English-style natural histories of their territories, without any success; only the works of Johann Jakob Scheuchzer in Switzerland, who effectively translated his Baconian enthusiasms into a German-speaking context, succeeded in establishing a model that Germans could and did eventually emulate. As this case study shows, differences in what might be called "national style" clearly did have significant impact on the frameworks within which investigations of local natural worlds might be carried out.[12]

Yet, equally clearly, "national styles" only barely begin to account for the range of differences that, as subsequent observers were to note, might exist between the covers of local natural-historical works, even those of the same genre. While local floras, regional mineralogies, and county natural histories (or German natural histories of individual territories) each came to develop their own distinctive set of conventions as to general format, differences on an even more local level persisted. From the very beginning of the seventeenth century, for example, compilers of local floras like Ludwig Jungermann in Altdorf drew eclectically on a wide range of nomenclatural and terminological possibilities left behind by Renaissance naturalists, using whichever they felt best suited their own circumstances. If a compiler liked one particular author's name for a plant, for example, he might elect to use it, while another compiler might end up using a completely different name.[13] This situation, confusing enough in itself, became increasingly aggravated during the later seventeenth century, as a wave of taxonomically-oriented botanists began to develop and propose their own new systems of nomenclature, which their fellow countrymen then often hastened to adopt and popularize. Thus, for example, late seventeenth- and early eighteenth-century English compilers began to follow the language of the systematically-inclined John Ray, while their French counterparts favored that of their own compatriot Joseph Pitton de Tournefort, and German botanists tended to opt for that of their own *Landsmann*, Augustus Quirinus Rivinus. In this case, local preferences seem to have evolved into *de facto* national styles.

Fundamentally, though, decisions about the creation of the "local" knowledge within local natural-historical works were, in fact, made at the local level. For the authors themselves were the ones who chose which particular combination of features to offer in their works: which nomenclature to follow, whose system to use, what kinds of information to include or,

[12] For a recent, though somewhat problematic account of the thorny issue of national styles in science, citing much of the literature to date on the topic, see John Henry, "National Styles in Science: A Possible Factor in the Scientific Revolution?", in *Geography and Revolution*, eds. David N. Livingstone and Charles W. J. Withers (Chicago, IL: University of Chicago Press, 2005), 43–74.

[13] One seventeenth-century botanist complained bitterly about the "nomenclature of German plants, which is burdened with incredible confusion": Johann Sigismund Elsholtz, *Flora Marchica, sive catalogus plantarum, quae partim in hortis electoralibus Marchiae Brandenburgicae primariis, Berolinensi, Aurangiburgico, & Potstamensi excoluntur: partim sua sponte passim proveniunt* (Berlin: ex officina Rungiana, 1663), 9. To make matters even worse, as one commentator on early local floras has noted, sometimes a single local flora might use *multiple* names for the same plant within the same catalogue: K. Wein, *Elias Tillandz's 'Catalogus plantarum' (1683) in Lichte seiner Zeit erklärt und gewürdigt: ein Beitrag zur Geschichte der Botanik in Finnland* (Helsinki: Finnische Literatur-Gesellschaft, 1930), 240.

alternatively, omit. In the case of the local flora, for example, many authors just seem to have had their own ideas about how they wanted their works to appear: whether with plants listed in particular ways, or with other pieces of information (such as the months in which plants flowered) left in or left out. Adding to the resulting differences between individual works was the fact that the kind of natural knowledge that local naturalists were recording was, of course, intensely place-bound. In other words, a plant that produced purple flowers in one area within Europe might in another area, if it were to be found there at all, produce blue or pink ones, owing to natural variation. Even if naturalists in both areas happened to be using exactly the same nomenclature, *then*, it might still be difficult, if not impossible, for them to verify that they were actually looking at the same species. The lack of a common nomenclature only made the situation worse. The different styles of local-flora writing that emerged in different places, then, especially with the use of discrepant systems of nomenclature, did indeed in some cases lead to substantial problems in the communication of natural knowledge. Linnaeus's critiques of local florists, discussed in the previous chapter, arose precisely for this reason: because he realized that in order to ensure that different accounts of nature's diversity would be commensurable, the diversity of the practices of local florists *themselves* would have to be brought under control.

This book has also sought to demonstrate, however, many of the ways in which the kinds of seemingly local knowledge created during this period were not, in fact, so "local" after all. Though the compilers of local natural inventories did indeed write based on an intimate familiarity with their surrounding environments, they did so, as we have seen, in large part out of an awareness of the *broader* worlds that lay beyond their towns and territories. For the culture of localism in which they participated was to a great degree a transnational one, with its small-town representatives eager to put their own locality on the map of the learned world, and to utilize the technologies of publication to increase its fame. They framed their accounts of "indigenous" natural kinds not *solipsistically*, but rather in relation to "exotic" counterparts. They eagerly read and discussed local natural histories of places near and far, and they corresponded and exchanged specimens with these works' authors, in the process *comparing* their own natural worlds with those to be found elsewhere. And as their confidence grew, as we have seen, they came to expand their scope well beyond anything we might today recognize as "local". From surveying small tracts a mere several miles in radius, they came to explore entire territories and even continents, moving far beyond the extent of any *one* individual's personal knowledge. In the process, they established new networks of collective empiricism.

In generating new ideas about natural variety and natural riches, the authors of these natural inventories helped to establish a place for "local" knowledge within the natural sciences – but at a price. For it no longer *was* truly "local knowledge" in the sense that we normally understand the term, namely in the sense of a knowledge that reflected the hard-won understandings held by the inhabitants of an area themselves. During the seventeenth and eighteenth centuries, one specific group within European society, namely learned physicians, took on the role of publishing inventories of local nature. They compiled not only local floras, but also regional mineralogies and natural histories of particular territories.[14] Regarding their duty as the study not only of the human body, but rather of "*physis,*" or nature in its entirety, these learned physicians took up during the seventeenth and eighteenth centuries the self-appointed task of compiling and making public the information they acquired about the local natural worlds of the places they lived in. Though scholars who have written on natural history have frequently presented the story of the discipline as one of gradual emancipation from medical dominance or even "tyranny,"[15] this is a model that the case of local natural history clearly does not fit. On the contrary, during the seventeenth and eighteenth centuries, not only did learned physicians remain involved with the pursuit of natural history, as they had been during the Renaissance, but they also became, in effect, spokespeople for *local* nature in particular. They were the ones who, in overwhelming numbers, published so many inventories of local natural kinds that even today, for example, as we saw in the Introduction, many parts of Europe are considered by botanists to be the most thoroughly "known" in the world.[16]

In the local inventories that physicians published, however, many previous forms of local natural knowledge – for example, on the medicinal uses of herbs, as well as on other kinds of uses and qualities of natural objects – did not survive the journey into print. Recent work has explored how, in the process of European expansion, early colonial venturers frequently relied on the assistance of indigenous peoples; in the process of codifying their findings, however, colonial writers often ended up excluding certain kinds of local knowledge, many of which may now be irrecoverable.[17] Yet this

---

[14] This was, of course, less the case in England than on the Continent, perhaps in part due to English traditions of mixing antiquarianism with local natural history; neither Robert Plot nor John Ray, for example, had any medical training.

[15] See for example Brian W. Ogilvie, *The Science of Describing: Natural History in Renaissance Europe* (Chicago, IL: University of Chicago Press, 2006), who, while convincingly arguing that sixteenth-century natural history saw the wide participation of large numbers of individuals who had not been medically trained, and that plant descriptions came during this period to be less and less focused on potential uses as *materia medica,* nonetheless repeatedly presents his results as the evidence of a move in natural history away from medicine (i.e. on pp. 29, 34, 48, 187), even while decrying such teleological narratives of "emancipation" from it (p. 38).

[16] See Figure 2, "Five-grade map of the state of world floristic knowledge as of 1979," to be found in the Introduction.

[17] See for example Londa Schiebinger, *Plants and Empire: Colonial Bioprospecting in the Atlantic World* (Cambridge, MA: Harvard University Press, 2004).

phenomenon seems to have taken place not only abroad, but within Europe itself as well. Take, for example, the case of the local flora. With its extremely brief and compact layout, the result of its origins as a pocket text for medical students, the local flora may be seen as having effectively eliminated entire layers of the local knowledge and acknowledgment of local informants commonly found in Renaissance herbals.

Thus, for example, as scholars have begun to attempt to recover earlier popular understandings of the uses of plants in Europe and elsewhere, in the hope of discovering new medicinal drugs, it has been necessary for them to base their attempts almost exclusively on the examination of entries in ancient, medieval, and Renaissance herbals; local floras would be useless for this purpose, because as we have seen, with very few exceptions, all such information was systematically excised from them at their time of publication.[18] Ironically enough, it was, of course, physicians who performed this excision of explicitly "medicinal" knowledge, and thus of a form of knowledge that, as historians of medicine have recently argued, in fact showed considerable similarities between its folk and learned versions.[19] Yet the loss of what we nowadays call "local knowledge" must nonetheless be seen as considerable. When Linnaeus subsequently condemned local florists for their own local variations in method, and proposed his own more rigorous strictures, yet another aspect of "local knowledge" may be seen as having been lost – namely that of the local florists themselves. An era may be seen as having come to an end.

The legacy of the inventories of nature discussed in this book, however, was to prove an enduring one. For Linnaeus and his followers did not eliminate the genres they attacked; they simply adapted them to meet their own purposes. By 1800, within a half-century after the arrival of the Linnaean system, most local floras and regional mineralogies had been rewritten to fit the new system. New ones, furthermore, continued to be generated regularly, often getting longer and even more detailed in the process.[20] Simultaneously, the state began to take a genuine interest in such works. While naturalists' efforts at princely patronage, in the years before Linnaeus, had been all-too-frequently rebuffed, funding now began to flow more securely. The latter half of the eighteenth century, for example, was to see the proliferation of "economic" and "technical" natural-historical works of all kinds. From "handbooks of economic plants" to "technical natural histories" of

---

[18] Bart K. Holland, ed., *Prospecting for Drugs in Ancient and Medieval European Texts: A Scientific Approach* (Amsterdam: Harwood Academic Publishers, 1996). Alain Touwaide of the Smithsonian Institution in Washington, DC is currently working on a computerized database of plant uses from Mediterranean antiquity through the Renaissance.

[19] See for example Mary Lindemann, *Health and Healing in Eighteenth-Century Germany* (Baltimore, MD: Johns Hopkins University Press, 1996), 373.

[20] For an example of this phenomenon, see G. F. W. Meyer's manuscript "Flora Göttingensis" from 1856, in four folio volumes (!). Universitätsbibliothek Göttingen, Hist nat 93b.

individual territories, these works drew equally from information previously collected (and still being collected) in local floras, regional mineralogies, and natural histories of specific territories.[21] It was in this context that the discipline of *Statistik* or "statistics" (literally, the study of the state) emerged in the late eighteenth century. Though at first based primarily on qualitative or verbal descriptions of individual states and their contents, both natural and civil, the discipline soon took on a quantitative or "statistical" cast, as inventories of territories became ever more methodical.[22] By the mid-nineteenth century, *Statistik* had become firmly entrenched as a tool of administration, both in Europe and in the colonies, with large-scale reshaping of entire landscapes, whether agricultural, forest, or other, often the result.[23]

Even as many local natural-historical works became subsumed under larger quantifying projects for the benefit of the state, however, local natural history retained its diversity of methods and purposes. The rising popularity of natural theology in the eighteenth century, for example, led an increasing number of clergymen throughout Europe to take up their pens in praise of local nature.[24] One of them was an English minister named Gilbert White,

[21] On the perceived utility of natural history during this period, see Henry A. Lowood, *Patriotism, Profit, and the Promotion of Science in the German Enlightenment: The Economic and Scientific Societies, 1760–1815* (New York: Garland, 1991), as well as the works on the history of cameralism cited in Chapter 3; Lisbet Koerner, "Women and Utility in Enlightenment Science," *Configurations* 2 (1995): 233–255; and E. C. Spary, "'Peaches Which the Patriarchs Lacked': Natural History, Natural Resources, and the Natural Economy in France," in *Oeconomies in the Age of Newton*, ed. Margaret Schabas and Neil De Marchi (Durham, NC: Duke University Press, 2003), 14–41.

[22] Mohammed Rassem and Justin Stagl, eds., *Statistik und Staatsbeschreibung in der Neuzeit, vornehmlich im 16.-18. Jahrhundert* (Paderborn: Schöningh, 1980); Sibylla Köhler, "Zur Sozialstatistik im Deutschland zwischen dem 18. und 20. Jahrhundert," Ph.D. thesis, Dresden, 1993; Peter Buck, "People Who Counted: Political Arithmetic in the Eighteenth Century," *Isis* 73 (1982): 28–45; Richard Olson, *The Emergence of the Social Sciences, 1642–1792* (New York: Twayne, 1993); and Andrea Rusnock, *Vital Accounts: Quantifying Health and Population in Eighteenth-Century England and France* (Cambridge: Cambridge University Press, 2002).

[23] David F. Lindenfeld, *The Practical Imagination: The German Sciences of State in the Nineteenth Century* (Chicago, IL: University of Chicago Press, 1997); and Marie-Noëlle Bourguet, *Déchiffrer la France: La statistique départementale à l'époque napoléonienne* (Paris: Gordon and Breach Science Publishers, 1988). For a colonial example, see Thomas Jefferson, *Notes on the State of Virginia* (Paris: n.n., 1784). On the use of the new quantitative techniques within forestry, see Henry E. Lowood, "The Calculating Forester: Quantification, Cameral Science, and the Emergence of Scientific Forestry Management in Germany", in *The Quantifying Spirit in the 18th Century*, ed. Tore Frängsmyr, J. L. Heilbron, and Robin E. Rider (Berkeley: University of California Press, 1990), 315–342.

[24] Sara Stebbins, *Maxima in minimis: Zum Empirie-und Autoritätsverständnis in der physikotheologischen Literatur der Frühaufklärung* (Frankfurt: Peter D. Lang, 1980). Within England, clerics had long written natural-historical works (see discussion of this point in Chapter 4), but outside of England, physicians had continued to form the majority of naturalist authors until the eighteenth century.

whose book *The Natural History and Antiquities of Selborne*, a lengthy and conversational work describing his pleasant ramblings through his peaceful village surrounds, was to become one of the best-known local natural histories of all time.[25] Though in many ways White's tome drew on the themes and concerns of a typical English county natural history, from its sociable tone to its indulgence in antiquities, it ended up attracting a much broader audience than its predecessors. Readers were charmed by his portrayal of the delights of natural-historical observation – which had already begun to receive quite favorable press from a range of eighteenth-century authors – and many decided to undertake the study of local nature themselves as a pleasing pastime.[26] *The Natural History and Antiquities of Selborne* would subsequently prove a major inspiration for the nineteenth- and twentieth-century development of "nature writing," which in turn would play a key role in sparking preservation movements and ultimately fully "environmental" ones.

Meanwhile, the geographic ambitions of naturalists were widening still further. Late in the eighteenth century, various individuals had begun to compile floras and natural histories of their own emerging nation-states,[27] and even to essay works covering entire continents.[28] In this they were greatly

[25] Gilbert White, *The Natural History and Antiquities of Selborne* (London: Printed by T. Bensley, 1789). For an excellent analysis of some of the localist origins of White's work, see Vladimir Jankovic, "The Place of Nature and the Nature of Place: The Chorographic Challenge to the History of British Provincial Science," *History of Science 38* (2000): 79–113.

[26] Among these people were women, who (unless they were married to or the daughters of physicians or other naturalists) had, during the early modern period, seldom found themselves encouraged to pursue interests in natural history beyond a mere interest in flowers. For a description of the new, female-friendly culture of natural history of this time, much of it localist, see Ann B. Shteir, *Cultivating Women, Cultivating Science: Flora's English Daughters and the Culture of Botany, 1760 to 1860* (Baltimore, MD: Johns Hopkins University Press, 1996). On further developments in local natural history, see David Elliston Allen, *The Naturalist in Britain: A Social History* (London: Penguin, 1978, c1976) and Denise Phillips, "Friends of Nature: Urban Sociability and Regional Natural History in Dresden, 1800–1850," *Osiris* 18 (2003): 43–59.

[27] See for example John Berkenhout, *Outlines of the Natural History of Great-Britain and Ireland* (London: P. Elmsly, 1789); Jan le Francq van Berkhey, *Natuurlyke historie van Holland* (Amsterdam: By Yntema en Tieboel, 1769–1811); and Johann Matthäus Bechstein, *Gemeinnützige Naturgeschichte Deutschlands nach allen drey Reichen* (Leipzig: bey Siegfried Lebrecht Crusius, 1793–1805). The English, exceptional as always, had in fact begun this trend somewhat earlier, with numerous eighteenth-century works (and even some seventeenth-century ones) claiming to survey "British" natural history in its entirety; however, it was only in the late eighteenth century that this began to be done with any thoroughness, in England as well as in continental Europe.

[28] In the European context, see for example Johann Nepomuk von Laicharding, *Vegetabilia europaea* (Innsbruck: prelo et sumtibus Joann. Thom. Nobil. de Trattern, 1790–1); and Johann August Ephraim Goeze, *Europäischer Fauna* (Leipzig: Weidmannsche Buchhandlung, 1803).

assisted by the systematic information-gathering that had begun to take place on a variety of levels, both inside Europe and beyond it. In the nineteenth-century United States, for example, the newly-founded nation poured much of its energies into undertaking natural surveys on a grand scale.[29] And as the imperial century proceeded, voyaging naturalists like Charles Darwin methodically collected and described thousands upon thousands of "new" species.[30] New disciplines and subdisciplines arose amidst this flood of information, among them biogeography, ecology, and biology itself.[31] In the process, they transformed even further the genres of local natural history that had, in so many ways, begun so small. With the recent advent of Geographic Information Systems (GIS) and other tools for gathering and organizing data on the natural world, biologists are now for the first time within serious reach of undertaking a final expansion of scale: the creation (in database form) of an inventory that would include every living species on this planet.[32] Should they ever achieve this goal – one naturalists have sought ever since the Renaissance encyclopedists – the time-honored efforts of the naturalist to understand the natural world in its seemingly infinite variety will have to find a new outlet.

This book has sought to show the diverse and local origins of attempts to create such natural inventories. When naturalists first set forth, in early modern Europe, to catalog the natural productions of their own "indigenous" and "domestic" surroundings, they did so amidst an expanding array of encounters with new worlds, whether with those of classical antiquity, of the New World itself, of the Old World, or simply of the natural worlds outside their doorsteps. In the process, they came to take inventory of nature in different ways, with different audiences and implied uses. The same situation confronts us today: as we draw on the tremendous quantities of information we have accumulated about the natural world, ultimately derived from

[29] George P. Merrill, ed., *Contributions to a History of American Geological and Natural History Surveys* (Washington: Government Printing Office, 1920); Margaret Welch, "Nature Writ Large and Small: Local, State, and National Natural Histories, 1850–1875," in *The Book of Nature: Natural History in the United States 1825–1875* (Boston, MA: Northeastern University Press, 1998), 90–132. See also David Spanagel, "Chronicles of a Land Etched by God, Water, Fire, Time, and Ice," Ph.D. thesis, Harvard University, 1996.

[30] Clare Lloyd, *The Travelling Naturalists* (London: Croom Helm, 1985); Janet Browne, *Charles Darwin: A Biography. Vol. I: Voyaging* (London: Pimlico, 1996).

[31] See Lynn Nyhart, "Natural History and the 'New' Biology," in *Cultures of Natural History*, ed. N. Jardine, J. A. Secord, and E. C. Spary (Cambridge: Cambridge University Press, 1996), 426–443.

[32] On recent efforts to quantify biodiversity and to assemble global databases of biological information, see David Takacs, *The Idea of Biodiversity: Philosophies of Paradise* (Baltimore, MD: Johns Hopkins University Press, 1996) and Geoffrey C. Bowker, "Databasing the World: Biodiversity and the 2000s," in *Memory Practices in the Sciences* (Cambridge, MA: The MIT Press, 2005), 107–138.

the contributions of numerous "local" informants around the globe, we can choose to use this information in a number of ways. We can use it, for example, to locate certain types of natural kinds, so that we can employ them to meet our needs. Or we can use it to attempt to protect these natural kinds. What will we do with our knowledge? That is up to us.

Works Cited

Manuscript Sources

Erlangen (Germany). Universitätsbibliothek. Handschriftenabteilung.
Trew-Sammlung.
Beringer correspondence. Cited by date of individual letter.
Brückmann correspondence. Cited by date of individual letter.
Ms. B248 (Johann Wilhelm Kretschmann, "Marchionatus Brandenburgico-Baruthini
Subterranei Specimen primum oder erster Versuch einer Beschreibung der Unterirdischen
Marggraffthums Brandenburg-Bayreuth")

Gotha (Germany). Forschungsbibliothek. Handschriftenabteilung.
Chart. B857b (Johann Philipp Breyne, drafts of correspondence)

Göttingen (Germany). Niedersächsische Staats- und Universitätsbibliothek. Abteilung
Handschriften und Alte Drucke.
Hist nat 93b (G. F. W. Meyer, "Flora Göttingensis")
Hist nat 96 (Anonymous, "Catalogus plantarum in Praefectura Wildershausen et
proxime adjacentibus sponte nascentium")
Hist nat 106 (Georg Forster, "Diarium faunae floraeque Vilnensis")

London (UK). British Library. Department of Manuscripts.
Add. 4436 (Johann Jakob Scheuchzer, "De terra Helvetica bonitate et qualitatibus
variantibus")
Sloane 3402 (Johann Jakob Scheuchzer, biographies of writers of natural history, letter C)
Sloane 4066, fol. 271r (Johann Philipp Breyne, letter to Hans Sloane)

Zürich (Switzerland). Zentralbibliothek. Handschriftenabteilung.
Ms. Z II 649 (Johann Jakob Scheuchzer, Stammbuch)
Ms. Z VIII 18 (Johann Jakob Scheuchzer, notes for new edition of "Bibliotheca
scriptorum historiae naturali omnium regionum inservientium")
Ms. Z VIII 601 (Johann Jakob Scheuchzer, early draft of "Historiae Helvetiae naturalis
prolegomena")
Ms. Z VIII 611 (Johann Jakob Scheuchzer, "Bibliotheca Botanica")

Printed Primary Sources

Agricola, Georg. *De re metallica*. Basel: apud Hieron Frobenium et Nicolaum Episcopium,
1556.
Albrecht, Benjamin Gottlieb. *Dissertatio inauguralis medica de aromatum exoticorum et nostratium praestantia*. Erfurt: typis Heringii, 1740.
Ammann, Paul. *Supellex botanica, hoc est: Enumeratio plantarum, quae non solum in horto
medico academiae Lipsiensis, sed etiam in aliis circa urbem viridariis, pratis ac sylvis &c.
progerminare solent*. Leipzig: sumptibus Joh. Christ. Tarnovii, 1675.

Anonymous. *The Libelle of Englyshe Polycye: A Poem on the Use of Sea-Power*, 1436. Ed. George Warner. Oxford: Clarendon Press, 1926.

Aristotle. *Parts of Animals*. Translated by E. S. Forster. Cambridge, MA: Harvard University Press, 1955.

Astruc, Jean. *Mémoires pour l'histoire naturelle de la province de Languedoc.* Paris: chez Guillaume Cavelier, 1740.

Bacon, Francis. *Sylva sylvarum.* London: Printed by J. H. for William Lee, 1627.

Baier, Johann Jakob. *Oryktographia Norica, sive rerum fossilium et ad minerale regnum pertinentium, in territorio Norimbergensi eiusque vicinia observatarum succincta descriptio.* Nuremberg: impensis Wolfgangi Michahellis, 1708.

———. *Horti medici Acad. Altorf. historia curiose conquisita.* Altdorf: typis Iod. Guil. Kohlesii, 1727.

Barba, Alvaro Alonso. *Arte de los Metales.* Madrid: en la Imprenta del Reyno, 1640.

Barclay, John. *Icon animorum.* London: apud Danielem & Davidem Aubrios, & Clementem Schleichium, 1614.

Bartholin, Thomas. *De medicina Danorum domestica dissertationes X.* Copenhagen: typis Matthiae Godicchenii, 1666.

Bauhin, Caspar. *Catalogus plantarum circa Basileam sponte nascentium cum earundem Synonymiis & locis in quibus reperiuntur: in usum Scholae Medicae, quae Basileae est.* Basel: typis Johan. Jacobi Genathii, 1622.

Becher, Johann Joachim. *Politische Discurs von den eigentlichen Ursachen dess Auff- und Abnehmens der Stadt, Lander, und Republicken.* Frankfurt am Main: in Verlegung Johann David Zunners, 1673.

Bechstein, Johann Matthäus. *Gemeinnützige Naturgeschichte Deutschlands nach allen drey Reichen.* Leipzig: bey Siegfried Lebrecht Crusius, 1793–1805.

Becmann, Johann Christoph. "Catalogus plantarum in tractu Francofurtano sponte nascentium." In *Memoranda Francofurtana: Notitia Universitatis, Catalogus Bibliothecae, Chronicon Civitatis, Catalogus Plantarum.* Frankfurt an der Oder: prelo Friderici Eichornii, 1676, 72–80.

Belleval, Pierre Richer de. *Dessein touchant la recherche des plantes du pays de Languedoc.* Montpellier: n.n., 1605.

———. *Opuscules.* Ed. M. Broussonet. Paris: n.n., 1785.

Beringer, Johann Bartholomäus Adam. *Lithographiae Wirceburgensis, ducentis lapidum figuratorum, a potiori insectiformium, prodigiosis imaginibus exornatae specimen primum.* Würzburg: apud Philippum Wilhelmum Fuggart, 1726.

———. *The Lying Stones of Dr. Johann Bartholomew Adam Beringer, being his Lithographiae Wirceburgensis.* Translated and annotated by Melvin E. Jahn and Daniel J. Woolf. Berkeley: University of California Press, 1963.

Berkenhout, John. *Outlines of the Natural History of Great-Britain and Ireland.* London: Printed for P. Elmsly, 1789.

Besler, Basil. *Hortus Eystettensis.* [Nuremberg]: n.n., 1613.

Beverwyck, Jan van. *Inleydinge tot de hollandse geneesmiddelen.* Dordrecht: voor Jasper Gorissz., 1642.

———. *Autarkeia Bataviae, sive introductio ad medicinam indigenam.* Leiden: ap. Joh. Maire, 1644.

Bidloo, Lambertus. *Dissertatio de re herbaria.* Amsterdam: apud H. & viduam T. Boom, 1683.

Blackstone, John. *Fasciculus plantarum circa Harefield sponte nascentium.* London: typis H. Woodfall, junioris, 1737.

Boate, Gerard. *Irelands Naturall History.* London: for John Wright, 1652.

Bock, Hieronymus. *New Kreütterbuch von Underscheydt, Würckung und Namen der Kreutter, so in Teutschen Lande wachsen.* Strasbourg: durch Wendel Rihel, 1539.

Bodin, Jean. *Methodus ad facilem historiarum cognitionem.* Paris: apud Martinum Iuvenem, 1566.

Borrichius, Olaus. *De usu plantarum indigenarum in medicina.* Copenhagen: literis & impensis Joh. Phil. Bockenhoffer, 1688.

Boyle, Robert. "General Heads for a Natural History of a Countrey, Great or Small." *Philosophical Transactions* 11 (April 1666): 186–189.

———. *General Heads for the Natural History of a Country, Great or Small; Drawn out for the Use of Travellers and Navigators.* London: Printed for John Taylor, 1692.

Bright, Timothie. *A Treatise wherein is declared the Sufficiencie of English Medicines, for cure of all diseases, cured with Medicine.* London: Printed by Henrie Middleton for Thomas Man, 1580.

Bromel, Olaus. *Chloris Gothica, seu Catalogus stirpium circa Gothoburgum nascentium. Det år ett Register uppå de Orter, Buskar och Träd som wåxa wilt kring om Gotheborg.* Göteborg: excudebat J. Rahm, 1694.

Browne, Sir Thomas. *Notes and Letters on the Natural History of Norfolk, More Especially on the Birds and Fishes.* London: Jarrold & Sons, 1902.

Brückmann, Franz Ernst. *Kurtze Beschreibung und genaue Untersuchung des Fürtrefflichen Weitzen-Biers, Duckstein genannt.* Braunschweig: n.n., 1723.

———. *Die Mumme scheut sich nicht/ sie will sich nicht verstecken.* Braunschweig: n.n., 1725.

———. *Thesaurus Subterraneus, Ducatus Brunsvigii, id est: Braunschweig mit seinen Unterirrdischen Schätzen und Seltenheiten der Natur.* Braunschweig: verlegt durch Johann Christoph Meisner, 1728.

———. *Ausführliche Beschreibung Einer seltsahmen Wunder-Geburt, Welche Eines abgedanckten Soldaten-Frau den 23. Febr. in Wolffenbüttel zur Welt gebracht.* Wolfenbüttel: bey Johann Christoph Meißner, 1732.

———. *Centuria epistolarum itinerarium.* Wolfenbüttel: n.n., 1742.

Brunfels, Otto. *Herbarum vivae eicones.* Strasbourg: apud Ioannem Schottum, 1530.

———. *Contrafayt Kreütterbüch.* Strasbourg: bey Hans Schotten, 1532.

Brunner, Johann. *Dissertatio physica de figuris variarum rerum in lapidibus & speciatim fossilibus comitatus Mansfeldici.* Leipzig: literis Joh. Georgi, 1675.

Bucher, Urban Gottfried. *Dissertationem medicam de catalepsi.* Wittenberg: ex Officina Goderitschiana, 1700.

———. *Dissertatio inauguralis medica leges naturae in corporum productione et conservatione.* Halle: J. Gruner, 1707.

———. *Zweyer guten Freunde vertrauter Brief-Wechsel vom Wesen der Seele.* The Hague: bey Peter von der Aa, 1713.

———. *Der Ursprung der Donau in der Landgraffschafft Fürstenberg, samt des Landes Beschaffen- und Vermögenheit, untersuchet, und mit andern hierzu dienenden Physicalischen Anmerckungen auch einigen Oeconomischen Reflexionen.* Nuremberg & Altdorf: bey Joh. Dan. Taubers sel. Erben, 1720.

———. *Das Muster eines Nützlich-Gelehrten in der Person Herrn Doctor Johann Joachim Bechers.* Nuremberg & Altdorf: bey Joh. Dan. Taubers seel. Erben, 1722a.

———. *Sachßen-Landes Natur-Historie, oder Beschreibung der Natürlichen Beschaffenheit und Vermögenheit der zu Sachßen gehörigen Provinzen.* Pirna: druckts Georg Balthasar Ludewig, 1722b.

———. *Sachsen-Landes Natur-Historie, In Welcher Dieses Landes, und der darzu gehörigen Provintzen Natürliche Beschaffenheit, Vermögenheit und Begebenheiten, in unterschiedenen Erzehlungen vorgestellet werden. Erste Erzehlung.* Dresden: bey Joh. Christoph Krausen, 1723.

Burton, Robert. *Anatomy of Melancholy.* London: Printed by John Lichfield and James Short, 1621.

Buxbaum, Johann Christian. *Enumeratio plantarum accuratior in agro Hallensi locisque vicinis crescentium.* Halle: in officina Libraria Rengeriana, 1721.

Calzolari, Francesco. *Il viaggio di Monte Baldo, della magnifica città di Verona.* Venice: Vincenzo Valgrisio, 1566.

Camerarius, Joachim. *Hortus medicus et philosophicus.* Frankfurt am Main: apud Hannem Feyerabend, 1588.

Carrichter, Bartholomäus. *Horn des Heyls menschlicher Blödigkeit. Oder, Kreütterbuch darinn die Kreütter des Teutschen Lands, ausz dem Liecht der Natur, nach rechter Art der himmelischen Einfliessungen beschriben.* Strasbourg: bey Christian Müller, 1576.

Champier, Symphorien. *Hortus gallicus, pro Gallis in Gallia scriptus.* Lyon: in aedibus Melchioris et Gasparis Trechsel fratrum, 1533.

Chemnitz, Johann. *Index plantarum circa Brunsvigam trium ferè milliarium circuitu nascentium.* Braunschweig: typis & sumtibus Christophori Friderici Zilligeri, 1652.

Childrey, Joshua. *Britannia Baconica: or, The Natural Rarities of England, Scotland, & Wales.* London: Printed for the Author, 1661.

Clusius, Carolus. *Rariorum aliquot stirpium per Hispanias observatarum historia.* Antwerp: ex officina Christophori Plantini, 1576.

———. *Rariorum aliquot stirpium per Pannoniam, Austriam, et vicinas . . . historia.* Antwerp: ex officina Christophori Plantini, 1583.

Commelin, Jan. *Nederlantze Hesperides, Dat is, Oeffening en Gebruik Van de Limoen- en Oranje-Boomen; Gestelt na den Aardt, en Climaat der Nederlanden.* Amsterdam: by Marcus Doornik, 1676.

———. *Catalogus plantarum indigenarum Hollandiae.* Amsterdam: apud H. & viduam T. Boom, 1683.

Cornut, Jacques Philippe. *Canadensium plantarum, aliarumque nondum editarum historia. Cui adiectum est ad calcem enchiridion botanicum Parisiense continens indicem plantarum, quae in pagis, silvis, pratis, & montosis iuxta Parisios locis nascuntur.* Paris: apud Simonem Le Moyne, 1635.

Culpeper, Nicholas. *The English Physitian: or, An Astrologo-Physical Discourse of the Vulgar Herbs of this Nation.* London: Printed by P. Cole, 1652.

Dapper, Olfert. *Exoticus curiosus.* Frankfurt am Main & Leipzig: bey Michael Rohrlachs seel. Wittib und Erben in Liegnitz, 1717.

Deering, Charles. *Catalogus stirpium &c., or, A catalogue of plants naturally growing and commonly cultivated in divers parts of England, more especially about Nottingham.* Nottingham: Printed for the Author by G. Ayscough, 1738.

Derham, William. *Physico-Theology.* London: Printed for W. Innys, 1713.

Dillenius, Johann Jakob. *Catalogus plantarum sponte circa Gissam nascentium.* Frankfurt am Main: apud J. Maximilianum à Sande, 1719.

———. *Hortus Elthamensis.* London: sumptibus auctoris, 1732.

———. *Historia muscorum.* Oxford: e theatro Sheldoniano, 1741.

Duvernoy, Johann Georg. *Designatio plantarum circa Tubingensem arcem florentium.* Tübingen: typis & impensis Georgii Friderici Pelickii, 1722.

Elsholtz, Johann Sigismund. *Flora Marchica, sive Catalogus Plantarum, Quae partim in hortis Electoralibus Marchiae Brandenburgicae primariis, Berolinensi, Aurangiburgico, & Potstamensi excoluntur: partim sua sponte passim proveniunt.* Berlin: ex officina Rungiana, 1663.

Fischer, Christian Gabriel. *Lapidum in agro Prussico sine praejudicio contemplandorum.* Königsberg: literis Reusnerianis, 1715.

Francke, Johannes. *Hortus Lusatiae.* Bautzen: excudebat Michael Wolrab, 1594.

Garidel, Pierre Joseph. *Histoire des plantes, qui naissent aux environs d'Aix et dans plusieurs autres endroits de la Provence.* Aix: J. David, 1715.

Gemeinhardt, Johann Caspar. *Catalogus plantarum circa Laubam nascentium, tam indigenarum, quam exoticarum.* Bautzen: apud Davidem Richterum, 1725.

Gesner, Conrad. *De omni rerum fossilium genere, gemmis, lapidibus, metallis, et huiusmodi.* Zürich: excudebat Iacobus Gesnerus, 1565.

Glauber, Johann Rudolf. *Teutschlands Wolfahrt.* Amsterdam: gedruckt bey Johan Jansson, 1656.

Goeze, Johann August Ephraim. *Europäischer Fauna.* Leipzig: Weidmannsche Buchhandlung, 1803.

Gottsched, Johann. See Loesel.

Grimm, Jacob and Wilhelm, eds. *Deutsches Wörterbuch.* Leipzig: S. Hirzel, 1854–1971.

Grisley, Gabriel. *Viridarium lusitanum.* Lisbon: ex praelo Antonii Craesbeeck, 1661.

Haller, Albrecht von. "Die Alpen." In *Versuch schweizerischer Gedichten.* Bern: Haller, 1732.

Helwing, Georg Andrea. *Flora quasimodogenita: sive enumeratio aliquot plantarum indigenarum in Prussia.* Danzig: imprimebat Joannes Daniel Stollius, 1712.

———. *Lithographia Angerburgica.* Königsberg: literis Johannis Stelteri, 1717.

Hengstmann, J. M. *Dissertatio medica inauguralis de medicamentis Germaniae indigenis sufficientibus.* Helmstedt: litteris Pauli Dieterici Schnorrii, 1730.

Hoffmann, Moritz. *Florae Altdorfinae deliciae hortenses sive Catalogus plantarum horti medici.* Altdorf: typis Georgi Hagen, 1660.

———. *Florae Altdorfinae deliciae sylvestres sive catalogus plantarum in agro Altdorffino locisque vicinis sponte nascentium.* Altdorf: typis Georgi Hagen, 1662.

———. *Botanotheca Laurembergiana.* Altdorf: typis Georgi Hagen, 1662.

———. *Montis Mauriciani in agro Leimburgensium, medio inter Norimbergam & Hirsbruccum.* Altdorf: typis Henrici Meyeri, 1694.

How, William. *Phytologia britannica, natales exhibens indigenarum stirpium sponte emergentium.* London: typis Ric. Cotes, 1650.

Jefferson, Thomas. *Notes on the State of Virginia.* Paris: n. n., 1784.

Jöcher, Christian Gottlieb. *Allgemeines Gelehrten-Lexicon.* Leipzig: J. F. Gleditsch, 1750.

Jockusch, Johann. *Versuch zur Natur-Historie der Graffschafft Mannßfeld.* Eisleben: gedruckt bey Joh. Philipp Hüllmann, 1730.

Johnson, Thomas. *Iter plantarum investigationis ergo susceptum a decem sociis, in agrum Cantianum.* [London?]: A. Mathewes, 1629.

Jungermann, Ludwig. *Catalogus plantarum, quae circa Altorfium Noricum, et vicinis quibusdam locis nascuntur.* Altdorf: apud Cunradum Agricolam, 1615.

———. *Catalogus plantarum, quae in horto medico Altdorfino reperiuntur.* Altdorf: typis Balthasari Scherffi, 1635.

———. *Catalogus plantarum, quae in horto medico et agro Altdorphino reperiuntur. Auctus & denuo recensitus.* Altdorf: typis Viduae Balthasari Scherffi, 1646.

Kentmann, Johannes. *Nomenclaturae rerum fossilium, quae in Misnia praecipue, & in aliis quoque regionibus inveniuntur.* Zürich: excudebat Jacobus Gesnerus, 1565.

Kentmann, Theophilus. *Tabula locum et tempus, quibus uberius plantae potissimum spontaneae vigent ac proveniunt, exprimens.* Wittenberg: apud Haeredes Samuelis Seelfisch, 1629.

Kestner, Christian Wilhelm. *Medicinisches Gelehrten-Lexicon.* Jena: bey Johann Meyers seel. Erben, 1740.

Knauth, Christoph. *Enumeratio plantarum circa Halam Saxonum.* Leipzig: sumpt. haered. Fried. Lanckisii, 1687.

Kretschmann, Johann Wilhelm. *Sammlung zu einer Berg-Historia des Marggrafthums Brandenburg-Bayreuth.* Ed. Dieter Arzberger. Selb-Oberweissenbach: Arzberger, 1994.

Kylling, Peder. *Viridarium danicum; sive, Catalogus trilinguis, latino-danico-germanicus plantarum indigenarum in Dania observatarum.* Copenhagen: n.n., 1688.

Lachmund, Friedrich. *Oryktographia Hildesheimensis, sive admirandorum fossilium, quae in tractu Hildesheimensi reperiuntur.* Hildesheim: sumptibus autoris, typis viduae Jacobi Mülleri, 1669.

Laicharding, Johann Nepomuk von. *Vegetabilia europaea*. Innsbruck: prelo et sumtibus Joann. Thom. Nobil. de Trattern, 1770–1771.

Lange, Carl Nicolaus. *Historia lapidum figuratorum Helvetiae*. Venice & Lucerne: sumptibus Authoris, typis Jacobi Tomasini, 1708.

Leigh, Charles. *The Natural History of Lancashire, Cheshire, and the Peak, in Derbyshire: with an Account of the British, Phoenician, Armenian, Gr. and Rom. Antiquities in those Parts.* Oxford: Printed for the Author, 1700.

Leopold, Johann Dietrich. *Deliciae silvestris florae Ulmensis*. Ulm: verlegts Johann Conrad Wohler, 1728.

Lerche, Johann Jacob. *Oryctographiam Halensem sive fossilium et mineralium in agro Halensi descriptionem*. Halle: typis Joh. Christiani Hilligeri, 1730.

Liebknecht, Johann Georg. *Hassiae Subterraneae specimen clarissima testimonia diluvii universalis*. Giessen & Frankfurt am Main: apud Eberh. Henr. Lammers, 1730.

Linnaeus, Carolus. *Bibliotheca botanica*. Amsterdam: apud Salomonem Schouten, 1736.

_____. *Oratio de necessitate peregrinationum intra patriam*. Leiden: apud Cornelium Haak, 1743.

_____. *The Lapland Journey*. Edited and translated by Peter Graves. Edinburgh: Lockharten Press, 1995.

Lister, Martin. *Historiae animalium Angliae*. London: apud Joh. Martyn, 1678.

Loesel, Johann. *Plantas in Borussia sponte nascentes*. Königsberg: typis Paschalii Mensenii, 1654.

_____. *Flora prussica, sive plantae in regno Prussiae sponte nascentes*. Revised and edited by Johann Gottsched. Königsberg: sumptibus Typographiae Georgianae, 1703.

Magnol, Pierre. *Botanicon Monspeliense*. Lyons: ex officina Francisci Carteron, 1676.

Martyn, John. *Methodus plantarum circa Cantabrigiam nascentium*. London: ex officina R. Reily, 1727.

Melle, Jakob von. *De lapidibus figuratis agri litorisque Lubecensis*. Lübeck: typis Struckianis, 1720.

Mentzel, Christian. *Centuria plantarum circa nobile Gedanum sponte nascentium*. Danzig: typis Andreae Hünefeldii, 1650. Reprinted in Reyger (see below).

Menzel, Albert. *Synonyma plantarum, seu simplicium, ut vocant, circa Ingolstadium sponte nascentium*. Ingolstadt: typis Ederianis, per Elisabetham Angermariam, viduam, 1618.

Mercati, Michele. *Metallotheca. Opus posthumum, auctoritate, & munificentia Clementis undecimi pontificis maximi e tenebris in lucem eductum; opera autem, & studio Joannis Mariae Lancisii, archiatri pontificii*. Rome: ex officina Jo: Mariae Salvioni, 1717.

Merrett, Christopher. *Pinax rerum naturalium Britannicarum, continens vegetabilia, animalia et fossilia in hac insula reperta inchoatus*. London: impensis Cave Pulleyn, typis F. & T. Warren, 1666.

Meyenberg, Heinrich Julius. *Flora Einbeccensis, sive enumeratio plantarum circa Einbeccam undique ad duo milliaria sponte nascentium*. Göttingen: literis Josquini Woyken, 1712.

Monti, Giuseppe. *Catalogi stirpium agri Bononiensis prodromus gramina ac hujusmodi affinia complectens*. Bologna: apud Constantinum Pisarri, 1719.

_____. *De monumento diluviano nuper in agro Bononiensi detecto*. Bologna: apud Rossi & Socios, 1719.

Morton, John. *The Natural History of Northampton-shire; with Some Account of the Antiquities. To which is Annex'd a Transcript of Doomsday-book, so far as it relates to That County.* London: Printed for R. Knaplock and R. Wilkin, 1712.

Mylius, Gottlob Friedrich. *Memorabilium Saxoniae subterraneae, i.e. Des Unterirdischen Sachsens seltsamer Wunder der Natur*. Leipzig: in Verlegung des Autoris, 1709–1718.

Oelhafen, Nicolaus. *Elenchus plantarum circa nobile Borussorum Dantiscum suâ sponte nascentium*. Danzig: typis & impensis Georgi Rheti, 1643.

Oldenburg, Henry. *The Correspondence of Henry Oldenburg*. Edited and translated by A. Rupert Hall and Marie Boas Hall. Madison: University of Wisconsin Press, 1965–1986.

Paracelsus. *Vom Holtz Guaiaco gründlicher Heylung.* Nuremberg: Friedrich Peypus, 1529.

———. *Von der frantzösischen Kranckheit.* Nuremberg: Friedrich Peypus, 1530.

———. "Herbarius Theophrasti de virtutibus herbarum, radicum seminum etc Alemaniae, patriae et imperii." In *Theophrast von Hohenheim gen. Paracelsus Sämtliche Werke. I. Abteilung: Medizinische, naturwissenschaftliche und philosophische Schriften.* Ed. Karl Sudhoff. Munich & Berlin: Oldenbourg, 1930, vol. II, 3–58.

———. *Sieben Defensiones, Verantwortung über etliche Verunglimpfungen seiner Missgönner.* In *Paracelsus: Four Treatises.* Edited and translated by Henry E. Sigerist. Baltimore, MD: Johns Hopkins University Press, 1941.

———. "The *Herbarius* of Paracelsus." Ed. Bruce Moran. *Pharmacy in History* 35 (1993): 99–127.

Paulli, Simon. *Flora danica; det er, Dansk urtebog.* Copenhagen: Aff Melchiore Martzen, 1648.

'Philanthropist'. "Substitutes for Tea." *The Pennsylvania Magazine; or, American Monthly Museum* (February 1775), 72–78.

Pilleter, Caspar. *Plantarum tum patriarum, tum exoticarum, in Walachria, Zeelandiae insula, nascentium synonymia.* Middelburg: excudebat Richardus Schilders, 1610.

Pitton, Jean-Scholastique. *De conscribenda Historia rerum Naturalium Provinciae.* Aix-en-Provence: apud Carolum David, 1672.

Pliny. *Natural History.* Edited and translated by T. E. Page et al. Cambridge, MA: Harvard University Press, 1938–1952.

Plot, Robert. *The Natural History of Oxford-shire, Being an Essay toward the Natural History of England.* Oxford: Printed at the Theater, 1677.

———. *The Natural History of Stafford-shire.* Oxford: Printed at the Theater, 1686.

Prévost, Jean. *Medicina pauperum.* Frankfurt am Main: sumptibus Johannis Beyeri, 1641.

Ray, John. *Catalogus plantarum circa Cantabrigiam nascentium.* Cambridge: excudebat J. Field, 1660.

———. *A Collection of English Proverbs.* Cambridge: Printed by J. Hayes, 1670.

———. *The Wisdom of God Manifested in the Works of the Creation.* London: Printed for Samuel Smith, 1691.

Reede tot Draakestein, Hendrik Adriaan van. *Hortus Indicus Malabaricus.* Amsterdam: sumptibus J. van Someren et J. van Dyck, 1678–1703.

Reyger, Gottfried. *Tentamen florae gedanensis methodo sexuali adcommodatae.* Danzig: apud Daniel Ludwig Wedel, 1764–1766.

Ritter, Albrecht. *Epistolica oryctographia Goslariensis.* Helmstedt: litteris Buchholzianis, 1732.

———. *Specimen I-II oryctographiae Calenbergicae.* Sondershausen: n.n., 1741–1743.

Robinson, Thomas. *An Essay towards a Natural History of Westmorland and Cumberland.* London: by W. Freeman, 1708.

Roggieri, Giovanni Giacomo. *Catalogo delle piante native del suolo Romano, co' loco principali Sinonimi, e luoghi natali.* Rome: per Felice Cesaretti, 1677.

Ruppe, Heinrich Bernhard. *Flora Jenensis.* Frankfurt am Main & Leipzig: apud Ernestum Claud. Bailliar., 1718.

Rzaczynski, Gabriel. *Historia naturalis curiosa regni Poloniae, magni ducatus Lituaniae, annexarumque provinciarum.* Sandomir: typis Collegii Soc. Jesu, 1721.

Sachs von Lewenhaimb, Philipp Jakob. *Oceanus macro-microcosmicus.* Breslau: sumptibus Esaiae Fellgiebelii, 1664.

Schaeffer, Karl. *Deliciae botanicae Hallenses.* Halle: typis Christophori Salfeldii, 1662.

Scheuchzer, Johann. *Agrostographiae Helveticae.* Zürich: sumptibus autoris, 1708.

Scheuchzer, Johann Jakob. *Charta invitatoria, quaestionibus, quae historiam Helvetiae naturalem concernunt, praefixa.* Zürich: n.n., 1699.

———. *Historiae Helveticae naturalis prolegomena.* Zürich: typis Davidis Gessneri, 1700.

———. *Beschreibung der Natur-Geschichten des Schweizerlandes.* Zürich: in Verlegung des Authoris, 1706–8.

———. *Piscium querelae et vindiciae.* Zürich: sumtibus Authoris, typis Gessnerianis, 1708.

———. *Bibliotheca Scriptorum Historiae Naturali omnium Terrae Regionum inservientium.* Zürich: typis Henrici Bodmeri, 1716.

———. *Helvetia historia naturalis, oder Natur-Historie des Schweitzerlandes.* Zürich: in der Bodmerischen Truckerey, 1716–1718.

———. *Homo diluvii testis et theoskopos.* Zürich: typis Joh. Henrici Byrgklini, 1726.

———. *Kupfer-Bibel.* Augsburg & Ulm: gedruckt bey Christian Ulrich Wagner, 1731–1735.

———. *Natur-Geschichte des Schweitzerlandes, samt seinen Reisen über die schweitzerische Gebürge.* Zürich: bey D. Gessner, 1746.

———. *Natur-Historie des Schweitzerlandes.* Zürich: bey Heidegger und Comp., 1752.

Schöpf, Johann. *Ulmischer Paradiss Garten, das ist: Eine Ferzeichnuss und Register, der Simplicien an der Zahl uber die 600. welche inn Gärten unnd nechstern Bezirck umb dess Statt Ulm zufinden.* Ulm: durch Johann Medern, 1622.

Schütte, Johann Heinrich. *Oryktographia Jenensis, sive fossilium et mineralium in agro Jenensi brevissima descriptio.* Leipzig: sumptibus Josephi Wolschendorfii, 1720.

Schwenckfeld, Caspar. *Stirpium & Fossilium Silesiae catalogus.* Leipzig: impensis Davidis Alberti, 1600.

———. *Theriotropheum Silesiae.* Liegnitz: impensis Davidis Alberti, 1603.

Seckendorff, Veit Ludwig von. *Teutscher Fürsten Stat.* Frankfurt am Main: in Verlegung Thomas Mattiae Gotzens, 1660.

Sloane, Hans. *Catalogus plantarum quae in insula Jamaica sponte proveniunt, vel vulgò coluntur, cum earundem synonymis & locis natalibus.* London: impensis D. Brown, 1696.

———. *A Voyage to the islands Madera, Barbados, Nieves, S. Christophers and Jamaica, with the Natural History of the Herbs and Trees, Four-footed Beasts, Fishes, Birds, Insects, Reptiles, &c. of the last of those Islands.* London: Printed by B. M. for the author, 1707–1725.

Strabo, Walahfrid. *Hortulus.* Translated by Raef Payne, with commentary by Wilfrid Blunt. Pittsburgh: The Hunt Botanical Library, 1966.

Thal, Johann. *Sylva Hercynia, sive catalogus plantarum sponte nascentium in montibus & locis plerisque Hercyniae Sylvae.* Frankfurt am Main: apud Iohannem Feyerabend, 1588.

Tillandz, Elias. *Catalogus plantarum, tam in excultis quam incultis locis prope Aboam superiori aestate nasci observatarum.* Åbo [Turku]: excusus â Johanne L. Wallio, 1673.

Tournefort, Joseph Pitton de. *Histoire des plantes qui naissent aux environs de Paris.* Paris: De l'Imprimerie Royale, 1698.

Valentini, Michael Bernhard. *Polychresta exotica.* Frankfurt: sumptibus Johannis Davidis Zunneri, 1700.

———. *Prodromus historiae naturalis Hassiae.* Giessen: typis Henningi Mülleri, 1707.

Volckamer, Johann Christoph. *Nürnbergische Hesperides, oder, Grundliche Beschreibung der edlen Citronat- Citronen- und Pomeranzen-Fruchte: wie solche in selbiger und benachbarten Gegend, recht mogen eingesetzt, gewartet, erhalten und fortgebracht werden.* Nuremberg: bei dem Authore, 1708–1714.

Volckamer, Johann Georg. *Flora Noribergensis, sive Catalogus plantarum in agro Noribergensi tam sponte nascentium, quam exoticarum.* Nuremberg: sumtibus Michaellianis, 1700.

Volkmann, Georg Anton. *Silesia subterranea, oder Schlesien mit seinen unterirdischen Schätzen.* Leipzig: verlegts Moritz Georg Weidmann, 1720.

Voltaire. *Philosophical Dictionary.* Edited and translated by Theodore Besterman. Harmondsworth, UK: Penguin, 1971.

Wagner, Johann Jakob. *Historia naturalis Helvetiae curiosa.* Zürich: impensis Joh. Henrici Lindinneri, 1680.

White, Gilbert. *The Natural History and Antiquities of Selborne.* London: Printed by T. Bensley, 1789.

Wigand, Johann. *Vera historia de succino Borussico, de alce Borussica, & de herbis in Borussia nascentibus*. Jena: typis Tobiae Steinmanni, 1590.

Will, Georg Andreas. *Nürnbergisches Gelehrten-Lexicon*. Nuremberg: zu finden bei L. Schupfel, 1755–58.

Wolfart, Peter. *Historiae naturalis Hassiae inferioris, Pars prima . . . i. e. Der Natur-Geschichte der Nieder-Furstenthums Hessen Erster Theil*. Kassel: gedruckt bey Heinrich Harmes, 1719.

Woodward, John. *Brief Instructions For Making Observations in all Parts of the World*. London: Printed for Richard Wilkin, 1696.

Secondary Sources

Agrimi, Jole and Chiara Crisciani. "Immagini e ruoli della 'vetula' tra sapere medico e antropologia religiosa (secoli XIII-XV)." In *Poteri carismatici e informali: chiesa e societa medioevali*. Ed. Jole Agrimi. Palermo: Sallerio, 1992, 224–261.

Aiton, E. J. *Leibniz: A Biography*. Bristol, UK: Adam Hilger, 1985.

Albala, Ken. *Eating Right in the Renaissance*. Berkeley: University of California Press, 2002.

Allen, D. E. "The Lost Limb: Geology and Natural History." In *Images of the Earth: Essays in the Environmental Sciences*. Ed. L. J. Jordanova and Roy S. Porter. Chalfont St. Giles, UK: British Society for the History of Science, 1979, 200–212.

——. "Walking the Swards: Medical Education and the Rise and Fall of the Botanical Field Class." *Archives of Natural History* 20 (1993): 335–367.

——. "Four Centuries of Local Flora-Writing: Some Milestones." *Watsonia* 24 (2003): 271–280.

Allut, Paul. *Étude biographique et bibliographique sur Symphorien Champier*. Lyon: Scheuring, 1859.

Alpers, Svetlana. *The Art of Describing: Dutch Art in the Seventeenth Century*. Chicago, IL: University of Chicago Press, 1983.

Álvarez Peláez, Raquel. *La conquista de la naturaleza americana*. Madrid: Consejo Superior de Investigaciones Científicas, 1993.

Ambrosoli, Mauro. *The Wild and the Sown: Agriculture and Botany in Western Europe, 1350–1850*. Cambridge: Cambridge University Press, 1997.

Anderson, Benedict. *Imagined Communities: Reflections on the Origin and Spread of Nationalism*. Revised ed. London: Verso, 1991.

Appadurai, Arjun, ed. *The Social Life of Things: Commodities in Cultural Perspective*. Cambridge: Cambridge University Press, 1996.

Arber, Agnes. *Herbals: Their Origin and Evolution: A Chapter in the History of Botany 1470–1670*. 3rd ed. Cambridge: Cambridge University Press, 1986.

Armitage, David and Michael J. Braddick, eds. *The British Atlantic World, 1500–1800*. Houndsmills, UK: Palgrave Macmillan, 2002.

Arrizabalaga, Jon, John Henderson, and Roger French. *The Great Pox: The French Disease in Renaissance Europe*. New Haven, CT: Yale University Press, 1997.

Ashworth, William B., Jr. "Natural History and the Emblematic World View." In *Reappraisals of the Scientific Revolution*. Ed. David C. Lindberg and Robert S. Westman. Cambridge: Cambridge University Press, 1990, 303–332.

Baader, Gerhard. "Medizinisches Reformdenken und Arabismus in Deutschland des 16. Jahrhunderts." *Sudhoffs Archiv* 63 (1979): 261–296.

Baehni, G. "Les grands systèmes botaniques depuis Linné." *Gesnerus* 14 (1957): 83–93.

Bailyn, Bernard. *Atlantic History: Concepts and Contours*. Cambridge, MA: Harvard University Press, 2005.

Baldwin, Martha. "Danish Medicines for the Danes and the Defense of Indigenous Medicines." In *Reading the Book of Nature: The Other Side of the Scientific Revolution*. Ed. Allen G.

Debus and Michael T. Walton. Kirksville, MO: Sixteenth Century Journal Publications, 1998, 163–180.

———. "Expanding the Therapeutic Canon: Learned Medicine Listens to Folk Medicine." In *Cultures of Communication from Reformation to Enlightenment: Constructing Publics in the Early Modern German Lands*. Ed. James Van Horn Melton. Aldershot, UK: Ashgate, 2002, 239–256.

Barnes, Sherman B. "The Editing of Early Learned Journals." *Osiris* 1 (1936): 155–172.

Barnett, Mason. "Medical Authority and Princely Patronage: The *Academia Naturae Curiosorum*, 1652–1693." Ph.D. thesis, University of North Carolina, Chapel Hill, 1995.

Baron, Hans. *The Crisis of the Early Italian Renaissance: Civic Humanism and Republican Liberty in an Age of Classicism and Tyranny*. Princeton, NJ: Princeton University Press, 1966.

Baron, W. "Gedanken über den ursprünglichen Sinn der Ausdrücke Botanik, Zoologie und Biologie." *Sudhoffs Archiv* 7 (1966): 1–10.

Barrera-Osorio, Antonio. *Experiencing Nature: The Spanish American Empire and the Early Scientific Revolution*. Austin: University of Texas Press, 2006.

Bayerl, Günter and Torsten Mayer. "Glueckseligkeit, Industrie und Natur-Wachstumsdenken im 18. Jahrhundert." In *Umweltgeschichte – Methoden, Themen, Potentiale*. Ed. Günter Bayerl, Norman Fuchsloch, and Torsten Meyer. Münster: Waxmann, 1996, 135–158.

Bender, Thomas, ed. *The University and the City: From Medieval Origins to the Present*. Oxford: Oxford University Press, 1988.

Bentley, Jerry. *Old World Encounters: Cross-Cultural Contacts and Exchanges in Pre-Modern Times*. Oxford: Oxford University Press, 1993.

Berg, Maxine and Helen Clifford, eds. *Consumers and Luxury: Consumer Culture in Europe 1650–1850*. Manchester: Manchester University Press, 1999.

Berry, Christopher J. *The Idea of Luxury: A Conceptual and Historical Investigation*. Cambridge: Cambridge University Press, 1994.

Biagioli, Mario. *Galileo, Courtier: The Practice of Science in the Culture of Absolutism*. Chicago, IL: University of Chicago Press, 1993.

———. and Peter Galison, eds. *Scientific Authorship: Credit and Intellectual Property in Science*. New York: Routledge, 2003.

Bicker, Alan, Paul Sillitoe, and Johan Pottier, eds. *Investigating Local Knowledge: New Directions, New Approaches*. Aldershot, UK: Ashgate, 2004.

Blair, Ann. "La persistance du latin comme langue de science à la fin de la Renaissance." In *Sciences et langues en Europe*. Ed. Roger Chartier and Pietro Corsi. Paris: École des Hautes Études en Sciences Sociales, 1996, 21–42.

———. *The Theater of Nature: Jean Bodin and Renaissance Science*. Princeton, NJ: Princeton University Press, 1997.

———. "Note Taking as an Act of Transmission." *Critical Inquiry* 31 (2004): 85–107.

Bleichmar, Daniela. "Books, Bodies, and Fields: Sixteenth-Century Transatlantic Encounters with New World *Materia Medica*." In *Colonial Botany: Science, Commerce, and Politics in the Early Modern World*. Ed. Londa Schiebinger and Claudia Swan. Philadelphia: University of Pennsylvania Press, 2005, 83–99.

Bog, Ingomar. *Der Reichsmerkantilismus. Studien zur Wirtschaftspolitik des Heiligen Römischen Reiches im 17. und 18. Jahrhundert*. Stuttgart: Fischer, 1959.

Bono, James. *The Word of God and the Languages of Man: Interpreting Nature in Early Modern Science and Medicine*. Madison: University of Wisconsin Press, 1995.

Bourguet, Marie-Noëlle. *Déchiffrer la France: La statistique départementale à l'époque napoléonienne*. Paris: Gordon and Breach Science Publishers, 1998.

Bowker, Geoffrey C. "Databasing the World: Biodiversity and the 2000s." In *Memory Practices in the Sciences*. Cambridge, MA: The MIT Press, 2005, 107–138.

Breen, T. H. *The Marketplace of Revolution: How Consumer Politics Shaped American Independence*. Oxford: Oxford University Press, 2004.

Brewer, John and Roy Porter, eds. *Consumption and the World of Goods*. London: Routledge, 1993.

Bright, Chris. *Life Out of Bounds: Bioinvasion in a Borderless World*. New York: W. W. Norton, 1998.

Brooke, John Hedley. *Science and Religion: Some Historical Perspectives*. Cambridge: Cambridge University Press, 1991.

Brotton, Jerry. *Trading Territories: Mapping the Early Modern World*. Ithaca, NY: Cornell University Press, 1997.

Browne, Janet. *The Secular Ark: Studies in the History of Biogeography*. New Haven, CT: Yale University Press, 1983.

———. *Charles Darwin: A Biography. Vol. I: Voyaging*. London: Pimlico, 1996.

Brückner, Jutta. *Staatswissenschaften, Kameralismus und Naturrecht*. Munich: Beck, 1977.

Bruford, Walter. *Germany in the Eighteenth Century: The Social Background of the Literary Revival*. Cambridge: Cambridge University Press, 1952.

Buck, Peter. "People Who Counted: Political Arithmetic in the Eighteenth Century." *Isis* 73 (1982): 28–45.

Buisseret, David, ed. *Monarchs, Ministers and Maps: The Emergence of Cartography as a Tool of Government in Early Modern Europe*. Chicago, IL: University of Chicago Press, 1992.

Burke, Peter. "*Heu domine, adsunt Turcae*: A Sketch for a Social History of Post-medieval Latin." In *Language, Self, and Society: A Social History of Language*. Ed. Peter Burke and Roy Porter. Cambridge: Polity Press, 1991, 23–50.

Burnby, J. "The Herb Women of the London Markets." *Pharmaceutical Historian* 13 (1983): 5–6.

Bylebyl, Jerome J. "The Medical Meaning of *Physica*." *Osiris* 6 (1990): 16–41.

Cameron, Rondo. "Economic Nationalism and Imperialism." In *A Concise Economic History of the World*. Oxford: Oxford University Press, 1993, 130–161.

Campbell, Mary B. *The Witness and the Other World: Exotic European Travel Writing, 400–1600*. Ithaca, NY: Cornell University Press, 1988.

Camporesi, Piero. "Retarded Knowledge." In *The Anatomy of the Senses: Natural Symbols in Medieval and Early Modern Italy*. Translated by Allan Cameron. Cambridge: Polity Press, 1994.

Cañizares-Esguerra, Jorge. "New World, New Stars: Patriotic Astrology and the Invention of Indian and Creole Bodies in Colonial Spanish America, 1600–1650." *American Historical Review* 104, 1 (1999): 33–68.

———. *How to Write the History of the New World: Histories, Epistemologies, and Identities in the Eighteenth-Century Atlantic World*. Stanford, CA: Stanford University Press, 2001.

Carney, Judith. *Black Rice: The African Origins of Rice Cultivation in the Americas*. Cambridge, MA: Harvard University Press, 2001.

Chaplin, Joyce. *Subject Matter: Technology, the Body, and Science on the Anglo-American Frontier, 1500–1676*. Cambridge, MA: Harvard University Press, 2001.

———. "Expansion and Exceptionalism in Early American History." *Journal of American History* 89 (2003): 1431–1455.

Clark, William. *Academic Charisma and the Origins of the Research University*. Chicago, IL: University of Chicago Press, 2006.

———. "The Scientific Revolution in the German Nations." In *The Scientific Revolution in National Context*. Ed. Roy Porter and Mikuláš Teich. Cambridge: Cambridge University Press, 1992, 90–114.

Coates, Ken S. *A Global History of Indigenous Peoples: Struggle and Survival*. Houndmills, UK: Palgrave Macmillan, 2004.

Coates, Peter. *American Perceptions of Immigrant and Invasive Species: Strangers on the Land.* Berkeley: University of California Press, forthcoming.

Cocchiara, Giuseppe. *The History of Folklore in Europe.* Philadelphia: Institute for the Study of Human Issues, 1952.

Coclanis, Peter A. "*Drang Nach Osten*: Bernard Bailyn, the World-Island, and the Idea of Atlantic History." *Journal of World History* 13 (2002): 169–182.

Coleman, D. C., ed. *Revisions in Mercantilism.* London: Methuen, 1969.

"Constructing Race: Differentiating Peoples in the Early Modern World", special issue of *William and Mary Quarterly,* 3rd ser., 54, 1 (1997).

Cook, Harold J. "Physick and Natural History in Seventeenth-Century England." In *Revolution and Continuity: Essays in the History and Philosophy of Early Modern Science.* Ed. Peter Barker and Roger Ariew. Washington, DC: Catholic University of America Press, 1991, 63–80.

———. "The Cutting Edge of a Revolution? Medicine and Natural History near the Shores of the North Sea." In *Renaissance and Revolution: Humanists, Scholars, Craftsmen and Natural Philosophers in Early Modern Europe.* Ed. J. V. Field and Frank A. J. L. James. Cambridge: Cambridge University Press, 1993, 45–61.

Cooper, Alix. "The Museum and the Book: The *Metallotheca* and the History of an Encyclopaedic Natural History in Early Modern Italy." *Journal of the History of Collections* 7, 1 (1995): 1–23.

———. "From the Alps to Egypt (and Back Again): Dolomieu, Scientific Voyaging, and the Construction of the Field in Eighteenth-Century Natural History." In *Making Space for Science: Territorial Themes in the History of Science.* Ed. Crosbie Smith and Jon Agar. London: Macmillan, 1998, 39–63.

———. "The Indigenous versus the Exotic: Debating Natural Origins in Early Modern Europe." *Landscape Research* 28, 1 (2003): 51–60 (in special issue on "The Native, Naturalized, and Exotic: Plants and Animals in History").

———. "'The Possibilities of the Land': The Inventory of 'Natural Riches' in the Early Modern German Territories." In *Oeconomies of Science in the Age of Newton.* Ed. Margaret Schabas and Neil DeMarchi. Durham, NC: Duke University Press, 2003, 139–153.

———. "Homes and Households." In *The Cambridge History of Science, Vol. 3: The Early Modern Period.* Ed. Katharine Park and Lorraine Daston. Cambridge: Cambridge University Press, 2006, 224–237.

———. "Latin Words, Vernacular Worlds: Language, Nature, and the 'Indigenous' in Early Modern Europe." *Journal of East Asian Science, Technology, and Medicine,* forthcoming (in special issue on "Global Science and Comparative History: Jesuits, Science, and Philology in China and Europe, 1550–1850").

Copenhaver, Brian. *Symphorien Champier and the Reception of the Occultist Tradition in Renaissance France.* The Hague: Mouton, 1979.

Cormack, Lesley B. "'Good Fences Make Good Neighbors': Geography as Self-Definition in Early Modern England." *Isis* 82 (1991): 639–661.

———. *Charting an Empire: Geography at the English Universities, 1580–1620.* Chicago, IL: University of Chicago Press, 1997.

Cosgrove, Denis. *Apollo's Eye: A Cartographic Genealogy of the Earth in the Western Imagination.* Baltimore, MD: Johns Hopkins University Press, 2001.

Courtwright, David T. *Forces of Habit: Drugs and the Making of the Modern World.* Cambridge, MA: Harvard University Press, 2001.

Crisciani, Chiara. "History, Novelty, and Progress in Scholastic Medicine." *Osiris* 6 (1990): 118–139.

Cronon, William. "A Place for Stories: Nature, History, and Narrative." *Journal of American History* 78 (1992): 1347–1376.

Crosby, Alfred W., Jr. *The Columbian Exchange: Biological and Cultural Consequences of 1492.* Westport, CT: Greenwood Press, 1972.

————. *Ecological Imperialism: The Biological Expansion of Europe, 900–1900*. Cambridge: Cambridge University Press, 1986.

Crowther-Heyck, Kathleen. "Wonderful Secrets of Nature: Natural Knowledge and Religious Piety in Reformation Germany." *Isis* 94 (2003): 253–273.

Dann, Otto. "Vom *Journal des Scavants* zur wissenschaftlichen Zeitschrift." In *Gelehrte Bücher vom Humanismus bis zur Gegenwart*. Ed. Bernhard Fabian and Paul Raabe. Wiesbaden: Harrassowitz, 1983, 63–80.

Dannenbauer, Heinz. *Die Entstehung des Territoriums der Reichstadt Nürnberg*. Stuttgart: Kohlhammer, 1928.

Daston, Lorraine. "Baconian Facts, Academic Civility, and the Prehistory of Objectivity." *Annals of Scholarship* 8 (1992): 337–363.

————. "How Nature Became the Other: Anthropomorphism and Anthropocentrism in Early Modern Natural Philosophy." In *Biology as Society, Society as Biology: Metaphors*. Ed. Sabine Maasen. Dordrecht: Kluwer, 1995, 37–56.

———— and Katherine Park. *Wonders and the Order of Nature, 1150–1750*. New York: Zone Books, 1998.

Davis, Natalie Zemon. "Proverbial Wisdom and Popular Errors." In *Society and Culture in Early Modern France*. Stanford, CA: Stanford University Press, 1975, 227–267.

————. "Beyond the Market: Books as Gifts in Sixteenth-Century France." *Transactions of the Royal Historical Society* 33 (1983): 69–88.

————. *Women on the Margins: Three Seventeenth-Century Lives*. Cambridge, MA: Harvard University Press, 1995.

Dear, Peter. "Space, Revolution, and Science." In *Geography and Revolution*. Ed. David N. Livingstone and Charles W. J. Withers. Chicago, IL: University of Chicago Press, 2005, 27–42.

DeJean, Joan. *Ancients Against Moderns: Culture Wars and the Making of a Fin de Siècle*. Chicago, IL: University of Chicago Press, 1998.

Delaporte, François. *Nature's Second Kingdom: Explorations of Vegetality in the Eighteenth Century*. Translated by Arthur Goldhammer. Cambridge, MA: The MIT Press, 1982.

Devine, Robert S. *Alien Invasion: America's Battle with Non-Native Animals and Plants*. Washington, DC: National Geographic Society, 1998.

Dickens, A. G. *The German Nation and Martin Luther*. London: Edward Arnold, 1974.

Diepgen, Paul. *Deutsche Volksmedizin: Wissenschaft, Heilkunde und Kultur*. Stuttgart: Enke, 1935.

Dilg, Peter. "Vom Ansehen der Arzneikunst: Historische Reflexionen in Kräuterbüchern des 16. Jahrhunderts." *Sudhoffs Archiv* 62, 1 (1978): 64–79.

Dittrich, Erhard. *Die deutschen und österreichischen Kameralisten*. Darmstadt: Wissenschaftliche Buchgesellschaft, 1973.

Drayton, Richard. *Nature's Government: Science, Imperial Britain, and the 'Improvement' of the World*. New Haven, CT: Yale University Press, 2000.

Eamon, William. *Science and the Secrets of Nature: Books of Secrets in Medieval and Early Modern Culture*. Princeton, NJ: Princeton University Press, 1994.

Eisenstein, Elizabeth. *The Printing Press as an Agent of Change: Communication and Cultural Transformations in Early Modern Europe*. Cambridge: Cambridge University Press, 1979.

Ellenberger, F. *Histoire de la géologie, II: 1660–1810*. Paris: Lavoisier, 1994.

Elliott, J. H. *The Old World and the New 1492–1650*. Cambridge: Cambridge University Press, 1970.

————. "Final Reflections: The Old World and the New Revisited." In *America in European Consciousness, 1493–1750*. Ed. Karen Ordahl Kupperman. Chapel Hill: University of North Carolina Press, 1995, 391–408.

Ernstberger, Anton. "Die feierliche Eröffnung der Universität Altdorf (29. Juni 1623)." In *Franken – Böhmen – Europa*. Kallmünz: Lassleben, 1959, 156–176.

Euler, Friedrich W. "Entstehung und Entwicklung deutscher Gelehrtengeschlechter." In *Universität und Gelehrtenstand 1400–1800*. Ed. Helmuth Rössler and Günther Franz. Limburg: Starke, 1970, 183–232.

Evans, R J. W. "Learned Societies in Germany in the Seventeenth Century." *European Studies Review* 7 (1977): 129–151.

Farber, Paul Lawrence. "Buffon and Daubenton: Divergent Traditions within the *Histoire naturelle*." *Isis* 66 (1975): 63–74.

Fauchier-Magnan, Adrien. *The Small German Courts in the Eighteenth Century*. Translated by Mervyn Savill. London: Methuen, 1958.

Febvre, Lucien. "*Frontière*: The Word and the Concept." In *A New Kind of History*. Edited by Peter Burke. Translated by K. Folca. London: Routledge & Kegan Paul, 1973, 208–218.

Findlen, Paula. "Jokes of Nature and Jokes of Knowledge: The Playfulness of Scientific Discourse in Early Modern Europe." *Renaissance Quarterly* 42 (1989): 292–331.

———. "The Economy of Scientific Exchange in Early Modern Italy." In *Patronage and Institutions: Science, Technology, and Medicine at the European Court, 1500–1750*. Ed. Bruce T. Moran. Woodbridge, UK: Boydell, 1991, 5–24.

———. *Possessing Nature: Museums, Collecting, and Scientific Culture in Early Modern Italy*. Berkeley: University of California Press, 1994.

———. "Francis Bacon and the Reform of Natural History in the Seventeenth Century." In *History and the Disciplines in Early Modern Europe*. Ed. Donald Kelley. Rochester, NY: University of Rochester Press, 1997, 239–260.

———. "The Scientist's Body: The Nature of a Woman Philosopher in Enlightenment Italy." In *The Faces of Nature in Enlightenment Europe*. Ed. Lorraine Daston and Gianna Pomata. Berlin: BWV, 2003, pp. 211–236.

Fischer, Frank. *Citizens, Experts, and the Environment: The Politics of Local Knowledge*. Durham, NC: Duke University Press, 2000.

Fischer, Hans. *Johann Jakob Scheuchzer (2. August 1672 – 23. Juni 1733), Naturforscher und Arzt*. Zürich: Leemann, 1973.

Flessa, Dorothee. "Die Professoren der Medizin zu Altdorf von 1580–1809." Diss. Med. thesis, Erlangen, 1969.

Foucault, Michel. *The Order of Things: An Archaeology of the Human Sciences*. New York: Vintage, 1973, c1970.

———. "Questions on Geography." In *Power-Knowledge: Selected Interviews and Other Writings 1972–1977*. Ed. Colin Gordon. Brighton, UK: Harvester, 1980, 63–77.

Fournier, Marian. "Enterprise in Botany: Van Reede and his Hortus Malabaricus." *Archives of Natural History* 14 (1987): 123–58, 297–338.

Franke, Hans. *Die Würzburger Lügensteine. Tatsachen, Meinungen und Lügengespinste uber eine der berühmtesten geologischen Spottfälschungen des 18. Jahrhunderts*. Würzburg: Schöningh, 1991.

Frasca-Spada, Marina and Nick Jardine, eds. *Books and the Sciences in History*. Cambridge: Cambridge University Press, 2000.

Fredrickson, George M. *Racism: A Short History*. Princeton, NJ: Princeton University Press, 2002.

Freedberg, David. *The Eye of the Lynx: Galileo, His Friends, and the Beginnings of Modern Natural History*. Chicago, IL: University of Chicago Press, 2002.

Freedman, Paul. *Images of the Medieval Peasant*. Stanford, CA: Stanford University Press, 1999.

French, Roger. *Ancient Natural History*. London: Routledge, 1994.

Frodin, D. G. *Guide to Standard Floras of the World*. 2nd ed. Cambridge: Cambridge University Press, 2001.

Fumagalli, Vito. *Landscapes of Fear: Perceptions of Nature and the City in the Middle Ages*. Translated by Shayne Mitchell. London: Polity Press, 1993.

Geertz, Clifford. *The Interpretation of Cultures*. New York: Basic Books, 1973.

———. *Local Knowledge: Further Essays in Interpretive Anthropology.* New York: Basic Books, 1983.

George, Wilma. *Animals and Maps.* Berkeley: University of California Press, 1969.

Gerbi, Antonello. *The Dispute of the New World: The History of a Polemic, 1750–1900.* Translated by Jeremy Moyle. Pittsburgh, PA: University of Pittsburgh Press, 1973.

———. *Nature in the New World: From Christopher Columbus to Gonzalo Fernandez de Oviedo.* Pittsburgh, PA: University of Pittsburgh Press, 1985.

Gerhard, Dietrich. "Regionalismus und ständisches Wesen als ein Grundthema europäischer Geschichte." *Historische Zeitschrift* 174 (1952): 303–37.

Glacken, Clarence J. *Traces on the Rhodian Shore: Nature and Culture in Western Thought from Ancient Times to the End of the Eighteenth Century.* Berkeley: University of California Press, 1967.

Gladstone, Jo. "'New World of English Words': John Ray, FRS, the Dialect Protagonist, in the Context of his Times (1658–1691)." In *Language, Self, and Society: A Social History of Language.* Ed. Peter Burke and Roy Porter. Cambridge: Polity Press, 1991, 115–153.

Gleeson, Janet. *The Arcanum.* New York: Warner Books, 1998.

Gohau, Gabriel. *Les sciences de la terre aux XVIIe et XVIIIe siècles: Naissance de la géologie.* Paris: Albin Michel, 1990.

Golinski, Jan. *Science as Public Culture: Chemistry and Enlightenment in Britain, 1760–1820.* Cambridge: Cambridge University Press, 1992.

Golowin, Sergius. *Die Weisen Frauen: Die Hexen und ihr Heilwissen.* Basel: Sphinx, 1982.

Goodman, Jordan. *Tobacco in History: The Cultures of Dependence.* London: Routledge, 1993.

Goody, Jack. "The Recipe, the Prescription, and the Experiment." In *The Domestication of the Savage Mind.* Cambridge: Cambridge University Press, 1977, 129–145.

Gould, Stephen Jay. *The Lying Stones of Marrakech: Penultimate Reflections on Natural History.* New York: Harmony Books, 2000.

Grafton, Anthony. "The World of the Polyhistors: Humanism and Encyclopedism." *Central European History* 18 (1985): 31–47.

———. *Defenders of the Text: The Traditions of Scholarship in an Age of Science, 1450–1800.* Cambridge, MA: Harvard University Press, 1991.

———. *New Worlds, Ancient Texts: The Power of Tradition and the Shock of Discovery.* Cambridge, MA: Harvard University Press, 1992.

———. *The Footnote: A Curious History.* Cambridge, MA: Harvard University Press, 1997.

———. "The Rest Versus the West." In *Bring Out Your Dead: The Past as Revelation.* Cambridge, MA: Harvard University Press, 2001, 77–93.

Graham, Loren R., Wolf Lepenies, and Peter Weingart, eds. *Functions and Uses of Disciplinary Histories.* Dordrecht: Reidel, 1983.

Gray, Marion W. *Productive Men, Reproductive Women: The Agrarian Household and the Emergence of Separate Spheres during the German Enlightenment.* New York: Berghahn, 2000.

Greenblatt, Stephen. *Marvelous Possessions: The Wonder of the New World.* Chicago, IL: University of Chicago Press, 1991.

Greengrass, Mark, Michael Leslie, and Timothy Raylor, eds. *Samuel Hartlib and Universal Reformation: Studies in Intellectual Communication.* Cambridge: Cambridge University Press, 1994.

Grieco, Allen J. "The Social Politics of Pre-Linnean Botanical Classification." *I Tatti Studies: Essays in the Renaissance* 4 (1992): 131–149.

Groening, Gert and Joachim Wolschke-Bulmahn. "Some Notes on the Mania for Native Plants in Germany." *Landscape Journal* 11, 2 (1992): 116–126.

Grove, Richard. *Green Imperialism: Science, Colonial Expansion and the Emergence of Global Environmentalism, 1660–1880.* Cambridge: Cambridge University Press, 1994.

Hall, Marcus and Peter Coates, eds. "The Native, Naturalized, and Exotic: Plants and Animals in Human History." Special issue of *Landscape Research* 28, 1 (2003).

Hall, Marie Boas. *Henry Oldenburg: Shaping the Royal Society*. Oxford: Oxford University Press, 2002.

Hannaford, Ivan. *Race: The History of an Idea in the West*. Baltimore, MD: Johns Hopkins University Press, 1996.

Hannaway, Caroline. "Environment and Miasmata." In *Companion Encyclopedia of the History of Medicine*. Ed. W. F. Bynum and Roy Porter. London: Routledge, 1993, 292–308.

Harris, Jonathan Gil. *Foreign Bodies and the Body Politic: Discourses of Social Pathology in Early Modern England*. Cambridge: Cambridge University Press, 1998.

Hay, Denys. *Europe: The Emergence of an Idea*. Edinburgh: Edinburgh University Press, 1957.

Headrick, Daniel R. *When Information Came of Age: Technologies of Knowledge in the Age of Reason and Revolution, 1700–1850*. Oxford: Oxford University Press, 2000.

Heater, Derek. *A Brief History of Citizenship*. New York: New York University Press, 2004.

Heckscher, Eli F. *Mercantilism*. Translated by Mendel Shapiro. New York: Garland Publishing, 1983, c1935.

Heesen, Anke te. *Der Weltkasten. Die Geschichte einer Bildenzyklopädie aus dem 18. Jahrhundert*. Göttingen: Wallstein, 1997.

———. *The World in a Box: The Story of an Eighteenth-Century Picture Encyclopedia*. Translated by Ann M. Hentschel. Chicago, IL: University of Chicago Press, 2002.

Heinsohn, Gunnar and Otto Steiger. *Die Vernichtung der Weisen Frauen*. Herbstein: März, 1985.

Hellyer, Marcus. "The Pocket Museum: Edward Lhwyd's *Lithophylacium*." *Archives of Natural History* 23 (1996): 43–60.

Heniger, Johann. *Hendrik Adriaan van Reede tot Drakenstein (1636–1691) and Hortus Malabaricus: A Contribution to the History of Dutch Colonial Botany*. Rotterdam: Balkema, 1986.

Henry, John. "National Styles in Science: A Possible Factor in the Scientific Revolution?" In *Geography and Revolution*. Ed. David N. Livingstone and Charles W. J. Withers. Chicago, IL: University of Chicago Press, 2005, 43–74.

Hoffmann, Julius. *Die Hausväterliteratur und die Predigten über den christlichen Hausstand*. Weinheim: Beltz, 1959.

Holborn, Hajo. *A History of Modern Germany*. 3 vols. Princeton, NJ: Princeton University Press, 1982.

Holland, Bart K., ed. *Prospecting for Drugs in Ancient and Medieval European Texts: A Scientific Approach*. Amsterdam: Harwood Academic Publishers, 1996.

Hunger, F. W. T. "Jan of Johannes Commelin." *Nederlandsch Kruidkundig Archief* 1924, 187–202.

Hunter, Michael. *Establishing the New Science: The Experience of the Early Royal Society*. Woodbridge, UK: Boydell, 1989.

———. "Robert Boyle and the Early Royal Society: A Reciprocal Exchange in the Making of Baconian Science." *British Journal for the History of Science*, forthcoming.

Hyde, Elizabeth. *Cultivated Power: Flowers, Culture, and Politics in the Reign of Louis XIV*. Philadelphia: University of Pennsylvania Press, 2005.

Impey, Oliver and Arthur MacGregor, eds. *The Origins of Museums: The Cabinet of Curiosities in Sixteenth- and Seventeenth-Century Europe*. Oxford: Clarendon Press, 1985.

Ivins, William. *Prints and Visual Communication*. Cambridge, MA: The MIT Press, 1967.

Jahn, Melvin E. "A Further Note on Dr. Johann Bartholomew Adam Beringer." *Journal of the Society for the Bibliography of Natural History* 4 (1963): 160–161.

———. "Some Notes on Dr. Johann Jakob Scheuchzer and on *Homo diluvii testis*." In *Toward a History of Geology*. Ed. Cecil J. Schneer. Cambridge, MA: The MIT Press, 1969, 192–213.

Jankovic, Vladimir. "The Place of Nature and the Nature of Place: The Chorographic Challenge to the History of British Provincial Science." *History of Science* 38 (2000): 79–113.

Jardine, Nicholas. *The Birth of History and Philosophy of Science: Kepler's 'A Defense of Tycho Against Ursus'*. Cambridge: Cambridge University Press, 1984.

———, James Secord, and Emma Spary, eds. *Cultures of Natural History*. Cambridge: Cambridge University Press, 1995.

Jaumann, Herbert. "Was ist ein Polyhistor?" *Studia Leibnitiana* 22 (1990): 76–89.

Johnson, Christine. "Bringing the World Home: Germany and the Age of Discovery." Ph.D. thesis, Johns Hopkins University, 2001.

Kaup, Martina. "Die Urbarmachung des Oderbruchs. Umwelthistorische Annäherung an ein bekanntes Thema." In *Umweltgeschichte – Methoden, Themen, Potentiale*. Ed. Günter Bayerl, Norman Fuchsloch, and Torsten Meyer. Münster: Waxmann, 1994, 111–133.

Kempe, Michael. *Wissenschaft, Theologie, Aufklarung: Johann Jakob Scheuchzer (1672–1733) und die Sintfluttheorie*. Epfendorf: Bibliotheca Academica, 2003.

Kenny, Neil. *Curiosity in Early Modern Europe: Word Histories*. Wiesbaden: Harrassowitz, 1998.

Keunecke, Hans-Otto, ed. *Hortus Eystettensis: Zur Geschichte eines Gartens und eines Buches*. Munich: Schirmer-Mosel, 1989.

Kirchner, Heinrich. "Die Würzburger Lügensteine im Lichte neuer archivalischer Funde." *Zeitschrift der Deutschen Geologischen Gesellschaft* 87 (1935): 607–615.

Koch, Manfred. *Geschichte und Entwicklung des bergmännischen Schrifttums*. Goslar: Hübener, 1963.

Koerner, Lisbet. "Women and Utility in Enlightenment Science." *Configurations* 2 (1995): 233–255.

———. "Daedalus Hyperboreus: Baltic Natural History and Mineralogy in the Enlightenment." In *The Sciences in Enlightened Europe*. Ed. William Clark, Jan Golinski, and Simon Schaffer. Chicago, IL: University of Chicago Press, 1999, 389–422.

———. *Linnaeus: Nature and Nation*. Cambridge, MA: Harvard University Press, 1999.

———. See also Rausing, Lisbet.

Köhler, Sybilla. "Zur Sozialstatistik im Deutschland zwischen dem 18. und 20. Jahrhundert." Ph.D. thesis, Dresden, 1993.

Kremers, Edward and George Urdang. *History of Pharmacy*. Ed. Glenn Sonnedecker. 4th ed. Madison, WI: American Institute of the History of Pharmacy, 1976.

Krempel, Ulla, ed. *Jan van Kessel d. Ä., 1626–1679. Die Vier Erdteile*. Munich: Alte Pinakothek, 1973.

Kronick, David A. *A History of Scientific and Technical Periodicals: The Origins and Development of the Scientific and Technical Press 1665–1790*. Metuchen, NJ: The Scarecrow Press, 1976.

Küster, Hansjörg. *Geschichte der Landschaft in Mitteleuropa: Von der Eiszeit bis zur Gegenwart*. Munich: Beck, 1995.

Kusukawa, Sachiko. *The Transformation of Natural Philosophy: The Case of Philip Melanchthon*. Cambridge: Cambridge University Press, 1995.

———. "Leonhart Fuchs on the Importance of Pictures." *Journal of the History of Ideas* 58 (1997): 403–427.

Laeven, A. H. *The 'Acta Eruditorum' under the Editorship of Otto Mencke (1644–1707): The History of an International Learned Journal Between 1682 and 1707*. Translated by Lynne Richards. Amsterdam: APA-Holland University Press, 1990.

Larson, James L. *Reason and Experience: The Representation of Natural Order in the Work of Carl von Linné*. Berkeley: University of California Press, 1971.

———. *Interpreting Nature: The Science of Living Form from Linnaeus to Kant*. Baltimore, MD: Johns Hopkins University Press, 1994.

Laudan, Rachel. *From Mineralogy to Geology: The Foundations of a Science, 1650–1830.* Chicago, IL: University of Chicago Press, 1987.

_____. "Histories of the Sciences and Their Uses: A Review to 1913." *History of Science* 31 (1993): 1–34.

Leiser, Wolfgang. "Das Landgebiet der Reichsstadt Nürnberg." In *Nürnberg und Bern: Zwei Reichsstädte und ihre Landgebiete.* Ed. Rudolf Endres. Erlangen: Universitätsbibliothek Erlangen, 1990, 227–260.

Lepenies, Wolf. *Das Ende der Naturgeschichte: Wandel kultureller Selbstverständlichkeiten in den Wissenschaften des 18. und 19. Jahrhunderts.* Munich: Hanser, 1976.

Levine, Joseph M. "Ancients and Moderns Reconsidered." *Eighteenth-Century Studies* 15 (1981): 72–89.

_____. "Natural History and the History of the Scientific Revolution." *Clio* 13 (1983): 57–73.

Lévi-Strauss, Claude. *Totemism.* Translated by Rodney Needham. Boston, MA: Beacon Press, 1963.

Liddell, Henry George and Robert Scott, eds. *A Greek-English Lexicon.* 9th ed. Oxford: Clarendon Press, 1951.

Lindemann, Mary. *Health and Healing in Eighteenth-Century Germany.* Baltimore, MD: Johns Hopkins University Press, 1996.

_____. *Medicine and Society in Early Modern Europe.* Cambridge: Cambridge University Press, 1999.

Lindenfeld, David F. *The Practical Imagination: The German Sciences of State in the Nineteenth Century.* Chicago, IL: University of Chicago Press, 1997.

Livingstone, David N. *Putting Science in its Place: Geographies of Scientific Knowledge.* Chicago, IL: University of Chicago Press, 2003.

_____ and Charles W. J. Withers, eds. *Geography and Revolution.* Chicago, IL: University of Chicago Press, 2005.

Lloyd, Clare. *The Travelling Naturalists.* London: Croom Helm, 1985.

Long, Pamela O. "The Openness of Knowledge: An Ideal and Its Context in 16th-Century Writing on Mining and Metallurgy." *Technology and Culture* 32 (1991): 318–355.

_____. *Openness, Secrecy, Authorship: Technical Arts and the Culture of Knowledge from Antiquity to the Renaissance.* Baltimore, MD: Johns Hopkins University Press, 2001.

Love, Harold. *Scribal Publication in Seventeenth-Century England.* Oxford: Oxford University Press, 1993.

Lovejoy, Arthur O. *The Great Chain of Being.* Cambridge, MA: Harvard University Press, 1936.

Lowood, Henry E. "The Calculating Forester: Quantification, Cameral Science, and the Emergence of Scientific Forestry Management in Germany." In *The Quantifying Spirit in the 18th Century.* Ed. Tore Frängsmyr, J. L. Heilbron, and Robin E. Rider. Berkeley: University of California Press, 1990, 315–342.

_____. *Patriotism, Profit, and the Promotion of Science in the German Enlightenment: The Economic and Scientific Societies, 1760–1815.* New York: Garland, 1991.

Mączak, Antoni. *Travel in Early Modern Europe.* Translated by Ursula Phillips. Cambridge: Polity Press, 1995.

Maffi, Luisa, ed. *On Biocultural Diversity: Linking Language, Knowledge, and the Environment.* Washington, DC: Smithsonian Institution Press, 2001.

Manilal, K. S., ed. *Botany and History of Hortus Malabaricus.* Rotterdam: Balkema, 1980.

Martí y Escayol, Maria Antònia. "Catalunya dins la xarxa científica de la illustració. John Polus Lecaan: medicina i botànica a Barcelona durant la Guerra de Successió." *Manuscrits* 19 (2001): 175–194.

Martin, Julian. *Francis Bacon, the State, and the Reform of Natural Philosophy.* Cambridge: Cambridge University Press, 1991.

Maschke, Erich, and Jürgen Sydow, eds. *Stadt und Universität im Mittelalter und in der frühen Neuzeit.* Sigmaringen: Thorbecke, 1977.

Mason, Peter. *Infelicities: Representations of the Exotic.* Baltimore, MD: Johns Hopkins University Press, 1998.

Mathee, Rudi. "Exotic Substances: The Introduction and Global Spread of Tobacco, Coffee, Cocoa, Tea, and Distilled Liquor, Sixteenth to Eighteenth Centuries." In *Drugs and Narcotics in History.* Ed. Roy Porter and Mikuláš Teich. Cambridge: Cambridge University Press, 1995, 24–51.

McKie, Douglas. "The Arrest and Imprisonment of Henry Oldenburg." *Notes and Records of the Royal Society of London* 6 (1948): 28–47.

McKitterick, David. *Print, Manuscript and the Search for Order.* Cambridge: Cambridge University Press, 2003.

Mendyk, Stan A. E. "Robert Plot: Britain's 'Genial Father of County Natural Histories.'" *Notes and Records of the Royal Society of London* 39 (1985): 159–177.

———. *'Speculum Britanniae': Regional Study, Antiquarianism, and Science in Britain to 1700.* Toronto: University of Toronto Press, 1989.

Merk, Walther. *Der Gedanke des gemeinen Besten in der deutschen Staats- und Rechtsentwicklung.* Darmstadt: Wissenschaftliche Buchgesellschaft, 1968, c1934.

Merrill, George P., ed. *Contributions to a History of American Geological and Natural History Surveys.* Washington: Government Printing Office, 1920.

Miller, David Philip and Peter Hanns Reill, eds. *Visions of Empire: Voyages, Botany, and Representations of Nature.* Cambridge: Cambridge University Press, 1996.

Miller, Genevieve. "'Airs, Waters, and Places' in History." *Journal of the History of Medicine and Allied Sciences* 17 (1962): 129–140.

Möbius, M. "Wie sind die Bezeichnungen Zoologie, Botanik und Mineralogie entstanden?" *Jenaische Zeitschrift für Medizin und Naturwissenschaft* 77 (1944), 216–229.

Moore, James. "Revolution of the Space Invaders: Darwin and Wallace on the Geography of Life." In *Geography and Revolution.* Ed. David N. Livingstone and Charles W. J. Withers. Chicago, IL: University of Chicago Press, 2005, 106–132.

Moran, Bruce. *The Alchemical World of the German Court: Occult Philosophy and Chemical Medicine of the Circle of Moritz of Hessen (1572–1632).* Stuttgart: Steiner, 1991.

———, ed. *Patronage and Institutions: Science, Technology and Medicine at the European Court, 1500–1750.* Woodbridge: Boydell, 1991.

———, ed. and trans. "The *Herbarius* of Paracelsus." *Pharmacy in History* 35 (1993): 99–127.

Morton, A. G. *History of Botanical Science: An Account of the Development of Botany from Ancient Times to the Present Day.* New York: Academic Press, 1981.

Muir, Edward. *Ritual in Early Modern Europe.* 2nd ed. Cambridge: Cambridge University Press, 2005.

Mukerji, Chandra. *Territorial Ambitions and the Gardens of Versailles.* Cambridge: Cambridge University Press, 1997.

———. "The Great Forestry Survey of 1669–1671: The Use of Archives for Political Reform." *Social Studies of Science,* forthcoming.

Müller-Wille, Staffan. *Botanik und weltweiter Handel. Zur Begründung eines natürlichen Systems der Pflanzen durch Carl von Linné (1707–1778).* Berlin: VWB, 1999.

Mulsow, Martin. "Säkularisierung der Seelenlehre? Biblizismus und Materialismus in Urban Gottfried Buchers *Briefwechsel vom Wesen der Seelen* (1713)." In *Säkularisierung in den Wissenschaften seit der frühen Neuzeit.* Ed. Sandra Pott and Lutz Danneberg. Berlin: Akademie-Verlag, 2002, 145–175.

Nazarea, Virginia D. *Cultural Memory and Biodiversity.* Tucson: University of Arizona Press, 1998.

Nettle, Daniel and Suzanne Romaine. *Vanishing Voices: The Extinction of the World's Languages.* Oxford: Oxford University Press, 2000.

Neumeister, Sebastian and Conrad Wiedemann, eds. *Res publica litteraria: Die Institutionen der Gelehrsamkeit in der frühen Neuzeit.* Wiesbaden: Harrassowitz, 1987.

Nicolson, Marjorie. *Mountain Gloom and Mountain Glory.* Ithaca: Cornell University Press, 1959.

Nielsen, Axel. *Die Entstehung der deutschen Kameralwissenschaft im 17. Jahrhundert.* Jena: Fischer, 1911.

Norton, Marcy. *Sacred Gifts, Profane Pleasures: A History of Tobacco and Chocolate.* Ithaca, NY: Cornell University Press, forthcoming.

Nutton, Vivian. "The Drug Trade in Antiquity." *Journal of the Royal Society of Medicine* 78 (1985): 138–145.

———, ed. *Medicine at the Courts of Europe, 1500–1837.* London: Routledge, 1990.

Nyhart, Lynn. "Natural History and the 'New' Biology." In *Cultures of Natural History.* Ed. N. Jardine, J. A. Secord, and E. C. Spary. Cambridge: Cambridge University Press, 1996, 426–443.

Ogilvie, Brian. *The Science of Describing: Natural History in Renaissance Europe.* Chicago, IL: University of Chicago Press, 2006.

Olmi, Giuseppe. "'Molti amici in varii luoghi': Studio della natura e rapporti epistolari nel secolo xvi." *Nuncius* 6 (1991): 3–31.

Olson, Richard. *The Emergence of the Social Sciences, 1642–1792.* New York: Twayne, 1993.

Olwig, Kenneth Robert. *Landscape, Nature, and the Body Politic: From Britain's Renaissance to America's New World.* Madison: University of Wisconsin Press, 2002.

Ong, Walter. "Latin Language Study as a Renaissance Puberty Rite." *Studies in Philology* 56 (1959): 103–124.

Ophir, Adi and Steven Shapin. "The Place of Knowledge: A Methodological Survey." *Science in Context* 4 (1991): 3–21.

Osborne, Michael A. "Acclimatizing the World: A History of the Paradigmatic Colonial Science." In *Nature and Empire: Science and the Colonial Enterprise.* Ed. Roy MacLeod. Chicago, IL: University of Chicago Press, 2000, 135–151.

Oslund, Karen. "'Nature in League with Man': Conceptualising and Transforming the Natural World in Eighteenth-Century Scandinavia." *Environment and History* 10 (2004): 305–325.

Outram, Dorinda. "New Spaces in Natural History." In *Cultures of Natural History.* Ed. N. Jardine, J. A. Secord, and E. C. Spary. Cambridge: Cambridge University Press, 1996, 249–265.

———. "On Being Perseus: New Knowledge, Dislocation, and Enlightenment Exploration." In *Geography and Enlightenment.* Ed. David N. Livingstone and Charles W. J. Withers. Chicago, IL: University of Chicago Press, 1999, 281–294.

Ozment, Steven. *Three Behaim Boys: Growing up in Early Modern Germany.* New Haven, CT: Yale University Press, 1990.

Pagden, Anthony. *European Encounters with the New World From Renaissance to Romanticism.* New Haven, CT: Yale University Press, 1993.

———, ed. *The Idea of Europe from Antiquity to the European Union.* Cambridge: Cambridge University Press, 2002.

Pagel, Walter. *Paracelsus: An Introduction to Philosophical Medicine in the Era of the Renaissance.* Basel: Karger, 1958.

Palmer, Richard. "Medical Botany in Northern Italy in the Renaissance." *Journal of the Royal Society of Medicine* 78 (1985): 149–157.

Park, Katharine. "The Meanings of Natural Diversity: Marco Polo on the 'Division' of the World." In *Texts and Contexts in Medieval Science: Studies on the Occasion of John E. Murdoch's Seventieth Birthday.* Ed. Edith Sylla and Michael R. McVaugh. Leiden: Brill, 1997, 134–147.

———. "Medicine and Magic: The Healing Arts." In *Gender and Society in Renaissance Italy.* Ed. Judith Brown and Robert Davis. London: Longman, 1998, 129–149.

———. "Natural Particulars: Medical Epistemology, Practice, and the Literature of Healing Springs." In *Natural Particulars: Nature and the Disciplines in Renaissance Europe*. Ed. Anthony Grafton and Nancy Siraisi. Cambridge, MA: The MIT Press, 1999, 347–367.

——— and Lorraine Daston. "Unnatural Conceptions: The Study of Monsters in Sixteenth- and Seventeenth-Century France and England." *Past and Present* 92 (1981): 20–54.

Patterson, Richard. "The 'Hortus Palatinus' at Heidelberg and the Reformation of the World." *Journal of Garden History* 1 (1981): 67–104 and 179–202.

Pelling, Margaret and Charles Webster. "Medical Practitioners." In *Health, Medicine and Mortality in the Sixteenth Century*. Ed. Charles Webster. Cambridge: Cambridge University Press, 1979, 165–235.

Perez-Ramos, Antonio. *Francis Bacon's Ideal of Science and the Maker's Knowledge Tradition*. Oxford: Clarendon Press, 1988.

Philip, Kavita. "Imperial Science Rescues a Tree: Global Botanic Networks, Local Knowledge, and the Transcontinental Transplantation of Cinchona." *Environment and History* 1 (1995): 173–200.

Phillips, Denise. "Friends of Nature: Urban Sociability and Regional Natural History in Dresden, 1800–1850." *Osiris* 18 (2003): 43–59.

Poeschel, Sabine. *Studien zur Ikonographie der Erdteile in der Kunst des 16.-18. Jahrhunderts*. Munich: Scaneg, 1985.

Pomata, Gianna, and Nancy Siraisi, eds. *Historia: Empiricism and Erudition in Early Modern Europe*. Cambridge, MA: The MIT Press, 2005.

Porter, Roy, ed. *The Medical History of Waters and Spas*. London: Wellcome Institute for the History of Medicine, 1990.

——— and Mikuláš Teich, eds. *Drugs and Narcotics in History*. Cambridge: Cambridge University Press, 1995.

Pratt, Mary Louise. *Imperial Eyes: Travel Writing and Transculturation*. London: Routledge, 1992.

Premuda, Loris. "Die Natio Germanica an der Universität Padua." *Sudhoffs Archiv* 47, 2 (1963): 97–105.

Prest, John. *The Garden of Eden: The Botanic Garden and the Re-Creation of Paradise*. New Haven, CT: Yale University Press, 1981.

Probst, Christian. *Fahrende Heiler und Heilmittelhändler. Medizin vom Marktplatz und Landstraße*. Rosenheim: Förg, 1992.

Quétel, Claude. *History of Syphilis*. Translated by Judith Braddock and Brian Pike. Baltimore, MD: Johns Hopkins University Press, 1990.

Ramati, Ayval. "Harmony at a Distance: Leibniz's Scientific Academies." *Isis* 87 (1996): 430–452.

Ranum, Orest. "Counter-Identities of Western European Nations in the Early-Modern Period: Definitions and Points of Departure." In *Concepts of National Identity: An Interdisciplinary Dialogue / Interdisziplinäre Betrachtungen zur Frage der nationalen Identität*. Ed. Peter Boerner. Baden-Baden: Nomos, 1986, 63–78.

Rappaport, Rhoda. *When Geologists Were Historians, 1665–1750*. Ithaca, NY: Cornell University Press, 1997.

Rassem, Mohammed and Justin Stagl, eds. *Statistik und Staatsbeschreibung in der Neuzeit, vornehmlich im 16.-18. Jahrhundert*. Paderborn: Schöningh, 1980.

Rausing, Lisbet. "Underwriting the Oeconomy: Linnaeus on Nature and Mind." In *Oeconomies in the Age of Newton*. Ed. Margaret Schabas and Neil De Marchi. Durham, NC: Duke University Press, 2003, 173–203.

———. See also Koerner, Lisbet.

Recktenwald, Horst. "Aufstieg und Niedergang der Universität Altdorf." In *Gelehrte der Universität Altdorf*. Ed. Horst Claus Recktenwald. Nuremberg: Spindler, 1966, 11–49.

Reeds, Karen Meier. "Renaissance Humanism and Botany." *Annals of Science* 33 (1976): 519–542.

_____. *Botany in Medieval and Renaissance Universities.* New York: Garland, 1991.

Reichenbach, Erwin and Georg Uschmann, eds. *Nunquam otiosus. Beiträge zur Geschichte der Präsidenten der Deutsche Akademie der Naturforscher Leopoldina.* Leipzig: Barth, 1970.

Reicke, Emil. *Geschichte der Reichsstadt Nürnberg von dem ersten urkundlichen Nachweis ihres Bestehens bis zu ihrem Übergang an das Königreich Bayern (1806).* Nuremberg: Raw, 1896.

Reill, Peter Hanns. *The German Enlightenment and the Rise of Historicism.* Berkeley: University of California Press, 1975.

Revel, Jacques. "Knowledge of the Territory." *Science in Context* 4 (1991): 133–161.

Ridder-Symoens, Hilde de, ed. *A History of the University in Europe, Volume 2: Universities in Early Modern Europe (1500–1800).* Cambridge: Cambridge University Press, 1995.

Riddle, John M. *Dioscorides on Pharmacy and Medicine.* Austin: University of Texas Press, 1985.

Ritvo, Harriet. "At the Edge of the Garden: Nature and Domestication in Eighteenth- and Nineteenth-Century Britain." *Huntington Library Quarterly* 55, 3 (1992): 363–378.

Roberts, R. S. "The Early History of the Import of Drugs into Britain." In *The Evolution of Pharmacy in Britain.* Ed. F. N. L. Poynter. London: Pitman Medical Publishing Company Ltd., 1965, 165–185.

Roessingh, H. K. "Tobacco Growing in Holland in the Seventeenth and Eighteenth Centuries: A Case Study of the Innovative Spirit of Dutch Peasants." *Acta Historiae Neerlandicae* 11 (1978): 18–54.

Röhrich, Heinz. "Zur Geschichte des 'Doctorgartens' oder 'Hortus medicus' der ehemaligen Nürnberger Universität Altdorf." *Erlanger Bausteine zur fränkischen Heimatforschung* 11 (1964): 31–42.

Rosenberg, Charles E. "Medical Text and Social Context: Explaining William Buchan's *Domestic Medicine.*" In *Explaining Epidemics and Other Studies in the History of Medicine.* Cambridge: Cambridge University Press, 1992, 32–56.

Rossi, Paolo. *Francis Bacon: From Magic to Science.* London: Routledge & Kegan Paul, 1968.

Rudwick, Martin J. S. *The Meaning of Fossils: Episodes in the History of Palaeontology.* Chicago, IL: University of Chicago Press, 1976.

_____. *The Great Devonian Controversy: The Shaping of Scientific Knowledge Among Gentlemanly Specialists.* Chicago, IL: University of Chicago Press, 1985.

Rusnock, Andrea. *Vital Accounts: Quantifying Health and Population in Eighteenth-Century England and France.* Cambridge: Cambridge University Press, 2002.

Sahlins, Peter. *Boundaries: The Making of France and Spain in the Pyrenees.* Berkeley: University of California Press, 1989.

Salaman, Redcliffe. *The History and Social Influence of the Potato.* Revised ed. Cambridge: Cambridge University Press, 1949.

Scarborough, John. "Roman Pharmacy and the Eastern Drug Trade: Some Problems as Illustrated by the Example of Aloe." *Pharmacy in History* 24 (1982): 135–143.

Schaffer, Simon. "The Earth's Fertility as a Social Fact in Early Modern Britain." In *Nature and Society in Historical Context.* Ed. Mikuláš Teich, Roy Porter, and Bo Gustafsson. Cambridge: Cambridge University Press, 1997, 124–147.

Schama, Simon. *The Embarrassment of Riches: An Interpretation of Dutch Culture in the Golden Age.* Berkeley: University of California Press, 1988.

_____. *Landscape and Memory.* New York: Alfred A. Knopf, 1995.

Scheel, Günter. "Leibniz und die deutsche Geschichtswissenschaft um 1700." In *Historische Forschung im 18. Jahrhundert.* Ed. Karl Hammer and Jürgen Voss. Bonn: Röhrscheid, 1976, 82–101.

Schiebinger, Londa. *The Mind Has No Sex? Women in the Origins of Modern Science.* Cambridge, MA: Harvard University Press, 1989.

————. *Nature's Body: Gender in the Making of Modern Science*. Boston, MA: Beacon Press, 1993.

———— *Plants and Empire: Colonial Bioprospecting in the Atlantic World*. Cambridge, MA: Harvard University Press, 2004.

————. and Claudia Swan, eds. *Colonial Botany: Science, Commerce, and Politics*. Philadelphia: University of Pennsylvania Press, 2005.

Schipperges, Heinrich. "Der Anti-Arabismus in Humanismus und Renaissance." In *Ideologie und Historiographie des Arabismus*. Wiesbaden: Franz Steiner Verlag, 1961, 14–25.

Schivelbusch, Wolfgang. *Tastes of Paradise: A Social History of Spices, Stimulants, and Intoxicants*. Translated by David Jacobson. New York: Vintage Books, 1992.

Schmid, Magnus. "Medici Universitatis Altorfinae." In *Gelehrte der Universität Altdorf*. Ed. Horst Claus Recktenwald. Nuremberg: Spindler, 1966, 79–98.

Schmidt, Benjamin. *Innocence Abroad: The Dutch Imagination and the New World, 1570–1670*. Cambridge: Cambridge University Press, 2001.

————. "Inventing Exoticism: The Project of Dutch Geography and the Marketing of the World, circa 1700." In *Merchants and Marvels: Commerce, Science, and Art in Early Modern Europe*. Ed. Pamela H. Smith and Paula Findlen. New York: Routledge, 2002, 347–369.

Schnalke, Thomas. *Medizin im Brief. Der städtische Arzt des 18. Jahrhunderts im Spiegel seiner Korrespondenz*. Stuttgart: Steiner, 1997.

Schöffer, I. "The Batavian Myth during the Sixteenth and Seventeenth Centuries." In *Britain and the Netherlands, V*. Ed. J. S. Bromley and E. H. Kossmann. The Hague: Nijhoff, 1975, 78–101.

Schorer, Edgar. "Der Autarkiebegriff im Wandel der Zeiten." *Jahrbuch für Gesetzgebung, Verwaltung und Volkswirtschaft im Deutschen Reich* 65 (1941): 47–82.

Schubart-Fikentscher, Gertrud. *Untersuchungen zur Autorschaft von Dissertationen im Zeitalter der Aufklärung*. Berlin: Akademie-Verlag, 1970.

Schwartz, Mark W. "Defining Indigenous Species: An Introduction." In *Assessment and Management of Plant Invasions*. Ed. James O. Luken and John W. Thieret. New York: Springer, 1997, 7–17.

Schwartz, Stuart B., ed. *Implicit Understandings: Observing, Reporting, and Reflecting on the Encounters Between Europeans and Other Peoples in the Early Modern Era*. Cambridge: Cambridge University Press, 1994.

Scott, James C. "Nature and Space." In *Seeing Like A State: How Certain Schemes to Improve the Human Condition Have Failed*. New Haven, CT: Yale University Press, 1998, 11–52.

Secord, Anne. "Artisan Botany." In *Cultures of Natural History*. Ed. N. Jardine, J. A. Secord, and E. C. Spary. Cambridge: Cambridge University Press, 1996, 378–393.

Seed, Patricia. *Ceremonies of Possession: Europe's Conquest of the New World, 1492–1640*. Cambridge: Cambridge University Press, 1996.

Sekora, John. *Luxury: The Concept in Western Thought*. Baltimore, MD: Johns Hopkins University Press, 1977.

Shapin, Steven. "The Invisible Technician." *American Scientist* 77 (October 1989): 554–563.

————. *A Social History of Truth: Civility and Science in Seventeenth-Century England*. Chicago, IL: University of Chicago Press, 1994.

————. "Placing the View from Nowhere: Historical and Sociological Problems in the Location of Science." *Transactions of the Institute of British Geographers*, 23 (1998), 5–12.

———— and Simon Schaffer. *Leviathan and the Air-Pump: Hobbes, Boyle, and the Experimental Life*. Princeton, NJ: Princeton University Press, 1985.

Shapiro, Barbara. "History and Natural History in Sixteenth- and Seventeenth-Century England: An Essay on the Relationship between Humanism and Science." In *English Scientific Virtuosi in the Sixteenth and Seventeenth Centuries*. Ed. Barbara Shapiro and Robert G. Frank, Jr. Los Angeles, CA: William Andrews Clark Memorial Library, 1979, 1–55.

Shteir, Ann B. *Cultivating Women, Cultivating Science: Flora's English Daughters and the Culture of Botany, 1760 to 1860.* Baltimore, MD: Johns Hopkins University Press, 1996.

Simpson, J. A. and E. S. C. Weiner, eds. *The Oxford English Dictionary.* Oxford: Oxford University Press, 1989.

Siraisi, Nancy G. *Avicenna in Renaissance Italy: The Canon and Medical Teaching in Italian Universities after 1500.* Princeton, NJ: Princeton University Press, 1987.

———. *Medieval and Early Renaissance Medicine: An Introduction to Knowledge and Practice.* Chicago, IL: University of Chicago Press, 1990.

Sloan, Phillip R. "Natural History, 1670–1802." In *Companion to the History of Modern Science.* Ed. R. C. Olby et al. London: Routledge, 1980, 295–313.

Small, Albion W. *The Cameralists: The Pioneers of German Social Policy.* Chicago, IL: University of Chicago Press, 1909.

Smith, Pamela H. *The Business of Alchemy: Science and Culture in the Holy Roman Empire.* Princeton, NJ: Princeton University Press, 1994.

———. *The Body of the Artisan: Art and Experience in the Scientific Revolution.* Chicago, IL: University of Chicago Press, 2004.

——— and Paula Findlen, eds., *Merchants and Marvels: Commerce, Science, and Art in Early Modern Europe.* New York: Routledge, 2002.

Smith, Wesley D. *The Hippocratic Tradition.* Ithaca, NY: Cornell University Press, 1979.

Spanagel, David. "Chronicles of a Land Etched by God, Water, Fire, Time, and Ice." Ph.D. thesis, Harvard University, 1996.

Spary, E. C. *Utopia's Garden: French Natural History from Old Regime to Revolution.* Chicago, IL: University of Chicago Press, 2000.

———. "'Peaches Which the Patriarchs Lacked': Natural History, Natural Resources, and the Natural Economy in France." In *Oeconomies in the Age of Newton.* Ed. Margaret Schabas and Neil De Marchi. Durham, NC: Duke University Press, 2003, 14–41.

Stafleu, Frans. "Botanical Gardens Before 1818." *Boissiera* 14 (1969): 31–46.

———. *Linnaeus and the Linnaeans: The Spreading of their Ideas in Systematic Botany, 1735–1789.* Utrecht: Oosthoek, 1971.

Stagl, Justin. "Vom Dialog zum Fragebogen: Miszellen zur Geschichte der Umfrage." *Kölner Zeitschrift für Soziologie und Sozialpsychologie* 31 (1979): 611–638.

———. *A History of Curiosity: The Theory of Travel, 1550–1800.* Chur, Switzerland: Harwood Academic Publishers, 1995.

Stannard, Jerry. "Natural History." In *Science in the Middle Ages.* Ed. David C. Lindberg. Chicago, IL: University of Chicago Press, 1978, 429–460.

Stearn, William T. *Botanical Latin.* 4th ed. London: David & Charles, 1992.

Stebbins, Sara. *Maxima in minimis: Zum Empirie- und Autoritätsverständnis in der physikotheologischen Literatur der Frühaufklärung.* Frankfurt: Peter D. Lang, 1980.

Steiger, Rudolf. *Johann Jakob Scheuchzer (1672–1733). 1. Werdezeit (bis 1699).* Zürich: Leemann, 1927.

Steinmeyer, Elias von. *Die Matrikel der Universität Altdorf.* Würzburg: Stürtz, 1912.

Stevens, Peter F. *The Development of Biological Systematics: Antoine-Laurent de Jussieu, Nature, and the Natural System.* New York: Columbia University Press, 1994.

Stiehler, G., ed. *Materialisten der Leibniz-Zeit.* Berlin: VEB Deutscher Verlag der Wissenschaften, 1966.

Stimson, Dorothy. "Hartlib, Haak and Oldenburg: Intelligencers." *Isis* 31 (1939–1940): 309–326.

Strauss, Gerald. *Sixteenth-Century Germany: Its Topography and Topographers.* Madison: University of Wisconsin Press, 1959.

Stroup, Alice. *A Company of Scientists: Botany, Patronage, and Community at the Parisian Royal Academy of Sciences.* Berkeley: University of California Press, 1990.

Sykora, K. V. "History of the Impact of Man on the Distribution of Plant Species." In *Biological Invasions in Europe and the Mediterranean Basin*. Ed. F. di Castri, A. J. Hansen, and M. Debussche. Dordrecht: Kluwer, 1990, 37–50.

Takacs, David. *The Idea of Biodiversity: Philosophies of Paradise*. Baltimore, MD: Johns Hopkins University Press, 1996.

Telle, Joachim, ed. *Pharmazie und der gemeine Mann*. Wolfenbüttel: Herzog August Bibliothek, 1982.

Temkin, Owsei. *Galenism: Rise and Decline of a Medical Philosophy*. Ithaca, NY: Cornell University Press, 1973.

Thomas, Keith. *Man and the Natural World: A History of the Modern Sensibility*. New York: Pantheon, 1983.

Todd, Kim. *Tinkering With Eden: A Natural History of Exotics in America*. New York: W. W. Norton, 2001.

Trevor-Roper, H. R. "Three Foreigners: The Philosophers of the Puritan Revolution." In *Religion, the Reformation, and Social Change and Other Essays*. London: Secker & Warburg, 1984, 237–293.

Troitzsch, Ulrich. *Ansätze technologischen Denkens bei den Kameralisten des 17. und 18. Jahrhunderts*. Berlin: Duncker & Humblot, 1966.

Valenčius, Conevery Bolton. "Histories of Medical Geography." In *Medical Geography in Historical Perspective*. Ed. Nicolaas Rupke. Amsterdam: Rodopi, 2000, 3–28.

Wakefield, R. Andre. "The Apostles of Good Police: Science, Cameralism, and the Culture of Administration in Central Europe, 1656–1800." Ph. D. thesis, University of Chicago, 1999.

Walker, Mack. *German Home Towns: Community, State, and General Estate 1648–1871*. Ithaca, NY: Cornell University Press, 1971.

Waquet, Françoise. *Latin, or the Empire of a Sign*. Translated by John Howe. London: Verso, 2001.

Wear, Andrew. "Making Sense of Health and the Environment in Early Modern England." In *Medicine and Society: Historical Essays*. Ed. Andrew Wear. Cambridge: Cambridge University Press, 1992, 119–147.

———. *Knowledge and Practice in English Medicine, 1550–1680*. Cambridge: Cambridge University Press, 2000.

———. "The Early Modern Debate about Foreign Drugs: Localism versus Universalism in Medicine." *The Lancet* 354 (July 10, 1999): 150.

Weaver, Jace. "Indigenousness and Indigeneity." In *A Companion to Postcolonial Studies*. Ed. Henry Schwarz and Sangeeta Ray. Oxford: Blackwell, 2000, 221–235.

Webster, Charles. *The Great Instauration: Science, Medicine and Reform 1626–1660*. New York: Holmes & Meier, 1973.

———. "Paracelsus: Medicine as Popular Protest." In *Medicine and the Reformation*. Ed. Ole Peter Grell and Andrew Cunningham. London: Routledge, 1993, 57–77.

Wein, Kurt. *Elias Tillandz's 'Catalogus plantarum' (1683) in Lichte seiner Zeit erklärt und gewürdigt. Ein Beitrag zur Geschichte der Botanik in Finnland*. Helsinki: Finnische Literatur-Gesellschaft, 1930.

———. "Die Wandlungen im Sinne des Wortes 'Flora'." *Fedde, Repertorium Specierum Novarum Regni Vegetabilis, Beihefte* 66 (1932): 74–87.

Welch, Margaret. "Nature Writ Large and Small: Local, State, and National Natural Histories, 1850–1875." In *The Book of Nature: Natural History in the United States 1825–1875*. Boston, MA: Northeastern University Press, 1998, 90–132.

Wijnands, D. O. *The Botany of the Commelins*. Amsterdam: Balkema, 1983.

Wilson, Kathleen. *The Sense of the People: Politics, Culture, and Imperialism in England, 1715–1785*. Cambridge: Cambridge University Press, 1998.

———. *The Island Race: Englishness, Empire, and Gender in the Eighteenth Century*. London: Routledge, 2002.

Wilson, Kevin and Jan van der Dussen, eds. *The History of the Idea of Europe*. London: Routledge, 1995.

Withers, Charles. *Geography, Science, and National Identity: Scotland since 1520*. Cambridge: Cambridge University Press, 2001.

Wolff, Larry. *Inventing Eastern Europe: The Map of Civilization on the Mind of the Enlightenment*. Stanford, CA: Stanford University Press, 1995.

Zantop, Susanne. *Colonial Fantasies: Conquest, Family, and Nation in Precolonial Germany, 1770–1870*. Durham, NC: Duke University Press, 1997.

Zart, Gustav. *Einfluss der englischen Philosophen seit Bacon auf die deutsche Philosophie des 18. Jahrhunderts*. Berlin: Dümmler, 1881.

Zaunick, Rudolph, Kurt Wein and Max Militzer, eds. *Johannes Franke 'Hortus Lusatiae' Bautzen 1594, mit einer Biographie*. Bautzen: Naturwissenschaftliche Gesellschaft Isis, 1930.

Zedelmaier, Helmut and Martin Mulsow, eds. *Die Praktiken der Gelehrsamkeit in der frühen Neuzeit*. Tübingen: Niemeyer, 2001.

Zielenziger, Kurt. *Die alten deutschen Kameralisten: Ein Beitrag zur Geschichte der Nationalökonomie und zum Problem des Merkantilismus*. Jena: Fischer, 1914.

Index